# Discourses on Sustaina
# Urban Mobility

# Discourses on Sustainable Urban Mobility

Robin Hickman

First published in 2025 by
UCL Press
University College London
Gower Street
London WC1E 6BT

Available to download free: www.uclpress.co.uk

Text © Author, 2025
Images © Author, 2025 unless otherwise stated in captions

The author has asserted his rights under the Copyright, Designs and Patents Act 1988 to be identified as the author of this work.

A CIP catalogue record for this book is available from The British Library.

Any third-party material in this book is not covered by the book's Creative Commons licence. Details of the copyright ownership and permitted use of third-party material is given in the image (or extract) credit lines. Every effort has been made to identify and contact copyright holders and any omission or error will be corrected if notification is made to the publisher. If you would like to reuse any third-party material not covered by the book's Creative Commons licence, you will need to obtain permission directly from the copyright owner.

This book is published under a Creative Commons Attribution-NonCommercial 4.0 International licence (CC BY-NC 4.0), https://creativecommons.org/licenses/by-nc/4.0/. This licence allows you to share and adapt the work for non-commercial use providing attribution is made to the author and publisher (but not in any way that suggests that they endorse you or your use of the work) and any changes are indicated. Attribution should include the following information:

Hickman, R. 2025. *Discourses on Sustainable Urban Mobility*. London: UCL Press. https://doi.org/10.14324/111.9781800089662

Further details about Creative Commons licences are available at https://creativecommons.org/licenses/

ISBN: 978-1-80008-964-8 (Hbk)
ISBN: 978-1-80008-965-5 (Pbk)
ISBN: 978-1-80008-966-2 (PDF)
ISBN: 978-1-80008-968-6 (epub)
DOI: https://doi.org/10.14324/111.9781800089662

Printed and bound by CPI Group (UK) Ltd, Croydon, CR0 4YY

# Contents

*List of figures* vii
*List of tables* xiii
*Foreword* xv
*Preface* xix
*Acknowledgements* xxi

**Part I: Introduction** 1

1  The production of mobility 3

2  Understanding space and mobility 15

**Part II: History** 43

3  The regimes and impacts of motorisation 45

4  Plymouth 55

**Part III: Truth** 71

5  Oxford 75

6  Freiburg 87

7  Singapore 99

8  Bogotá 111

9  Houten 123

**Part IV: Power** — 133

10  Chongqing — 135

11  London, King's Cross — 145

12  Rio de Janeiro — 157

**Part V: Discontinuity** — 165

13  Utrecht — 169

14  Copenhagen — 179

15  Malmö — 189

16  Dar es Salaam — 199

17  Delhi — 207

18  Shenzhen — 215

**Part VI: Ethics** — 221

19  Manchester — 225

20  Valenciennes — 239

21  Medellín — 245

**Part VII: Subjectivity** — 255

22  London, LTN21 — 257

23  Portland — 269

**Part VIII: Conclusion** — 289

24  Emerging mobility and space? — 291

*References* — 323
*Index* — 335

# List of figures

| | | |
|---|---|---|
| 0.1 | Illustration from 'La Dioptrique' by Rene Descartes (1596–1650) | xxiii |
| 2.1 | Visioning and backcasting | 23 |
| 2.2 | A framework for sustainable urban mobility | 25 |
| 2.3 | Key Foucauldian concepts in discourse | 31 |
| 2.4 | The motility concept | 36 |
| 2.5 | The spatial triad in transport planning | 37 |
| 2.6 | Case studies | 40 |
| 3.1 | Motorisation rates by country | 47 |
| 3.2 | The changing shape of car advertising | 50 |
| 3.2a | Baker Electrics, 1910s | 50 |
| 3.2b | Ford Mustang, 1960s | 50 |
| 3.2c | Toyota Corolla, 1970s | 50 |
| 3.2d | Land Rover Discovery, 2019 | 51 |
| 3.2e | Toyota Prius, 2010 | 51 |
| 3.2f | Audi Q7, 2016 | 51 |
| 4.1 | Plymouth city centre and surrounding developments | 57 |
| 4.2 | Plymouth motorisation and the city | 59 |
| 4.3 | Plymouth highway planning | 60 |
| 4.4 | Charles Church and highway dominance | 60 |
| 4.5 | Pedestrianisation in the city centre | 61 |
| 4.6 | 'Sustainable neighbourhood' at Sherford | 61 |
| 4.7 | Poor cycling provision in Sherford | 62 |
| 5.1 | The southern relief road proposals for Oxford | 76 |
| 5.2 | Transport and streetscape interventions in Oxford | 78 |
| 5.3 | Oxford and the bicycle | 79 |
| 5.4 | Broad Street | 79 |
| 5.5 | Christ Church meadow | 80 |
| 5.6 | Bus gate on High Street | 80 |
| 5.7 | Queen Street and semi-pedestrianisation | 81 |
| 5.8 | Frideswide Square and shared space | 81 |

| | | | |
|---|---|---|---|
| 5.9 | Oxford railway station | | 82 |
| 5.10 | Woodstock and car dominance | | 82 |
| 6.1 | TDM in the city centre of Freiburg | | 89 |
| 6.2 | Freiburg city centre | | 90 |
| 6.3 | Blaue Brücke and active mobility | | 91 |
| 6.4 | Tramway and the Vauban neighbourhood | | 91 |
| 6.5 | Community space in Vauban | | 92 |
| 6.6 | Shared garden areas | | 92 |
| 6.7 | Rieselfeld neighbourhood | | 93 |
| 6.8 | Car parking at the rear of buildings | | 93 |
| 6.9 | Cycle facilities | | 94 |
| 6.10 | Compact urban planning | | 94 |
| 6.11 | Vauban and Rieselfeld neighbourhood extensions in Freiburg | | 95 |
| 7.1 | Singapore MRT and the proposed North–South Corridor | | 103 |
| 7.2 | Singapore Marina Bay | | 105 |
| 7.3 | Electronic road pricing | | 106 |
| 7.4 | The extensive MRT system | | 106 |
| 7.5 | Toa Payoh new town | | 107 |
| 7.6 | Footway provision | | 107 |
| 7.7 | Chinatown | | 108 |
| 7.8 | The North–South Corridor | | 109 |
| 7.9 | Highway provision | | 109 |
| 8.1 | The TransMilenio network in Bogotá | | 113 |
| 8.2 | Highway provision | | 117 |
| 8.3 | The TransMilenio network | | 117 |
| 8.4 | City centre public spaces | | 118 |
| 8.5 | Cycle provision | | 118 |
| 8.6 | Alameda El Porvenir cycle route | | 119 |
| 8.7 | The Ciclovía and active travel | | 119 |
| 8.8 | Cable car access | | 120 |
| 8.9 | Bogotá's Ciclovía | | 121 |
| 9.1 | Houten new town | | 125 |
| 9.2 | Houten town centre | | 128 |
| 9.3 | Houten railway station | | 129 |
| 9.4 | The segregated cycle network | | 129 |
| 9.5 | Cycle parking | | 130 |
| 9.6 | Highway access to residential areas | | 130 |
| 9.7 | Traffic filtering | | 131 |
| 9.8 | More recent neighbourhoods | | 131 |
| 9.9 | Cycling boardwalk | | 132 |

| | | |
|---|---|---|
| 10.1 | Current and planned Chongqing Metro network by 2035 | 137 |
| 10.2 | Extensive Metro network | 139 |
| 10.3 | Integrated urban planning and transport infrastructure at Liziba station | 140 |
| 10.4 | Chongqing HSR, part of the national network | 140 |
| 10.5 | An extensive HSR station in Chongqing | 141 |
| 10.6 | HSR and conventional railways in China | 142 |
| 11.1 | King's Cross station area redevelopment | 147 |
| 11.2 | King's Cross Western Concourse | 149 |
| 11.3 | Granary Square and Central St Martins | 150 |
| 11.4 | Coal Drops Yard | 150 |
| 11.5 | Residential development at the Gasholders | 151 |
| 11.6 | Office development, landscaping and the view into the city centre | 152 |
| 12.1 | Transport and streetscape interventions at Praça Mauá and Rio Branco, Rio de Janeiro | 159 |
| 12.2 | The BRT system in Rio | 161 |
| 12.3 | Praça Mauá renewal | 161 |
| 12.4 | Porto Maravilha redevelopment | 162 |
| 12.5 | New tramway in Porto Maravilha | 162 |
| 12.6 | Morro da Providência | 163 |
| 13.1 | New cycling facilities and public spaces in Utrecht | 171 |
| 13.2 | Central Station in Utrecht | 173 |
| 13.3 | Rail connections are provided between urban centres | 174 |
| 13.4 | High-quality cycle parking | 174 |
| 13.5 | Cycle path and linear park at the Croeselaan | 175 |
| 13.6 | The city canal is reinstated at Catharijnesingel | 175 |
| 13.7 | Extensive cycle facilities | 176 |
| 13.8 | Dafne Schippers bridge and cycle path over the school | 176 |
| 13.9 | Leidsche Rijn new neighbourhood | 177 |
| 13.10 | High-quality cycling experience | 177 |
| 14.1 | New urban development around the public transport network in Copenhagen | 182 |
| 14.2 | The Nørrebrogade and C95 cycle superhighway | 183 |
| 14.3 | The modern practice of cycling | 184 |
| 14.4 | Lille Langebro cycle bridge | 185 |
| 14.5 | Cykelslangen bridge | 186 |
| 14.6 | Ørestad neighbourhood | 187 |
| 14.7 | Residential development in Ørestad | 187 |
| 14.8 | Communal open space in residential areas | 188 |
| 15.1 | Bo01 and Västra Hamnen | 190 |

| 15.2 | The Western Harbour redevelopment | 193 |
| 15.3 | 'European Village' | 193 |
| 15.4 | Spaces between the buildings | 194 |
| 15.5 | Streets are designed to restrict and slow traffic | 195 |
| 15.6 | Secure cycle storage | 196 |
| 15.7 | Stapelbäddsparken urban skatepark | 197 |
| 15.8 | Multi-storey car parking is hidden | 197 |
| 15.9 | Central railway station | 198 |
| 16.1 | The DART system in Dar es Salaam | 201 |
| 16.2 | Emerging motorisation in Dar es Salaam | 202 |
| 16.3 | The DART system | 202 |
| 16.4 | Modern bus fleet | 203 |
| 16.5 | Bus station and waiting environment | 203 |
| 16.6 | Segregated bus routes | 204 |
| 16.7 | Cycle lanes alongside the bus lanes | 204 |
| 16.8 | Public transport in the city centre | 205 |
| 17.1 | The Delhi Metro, Delhi-Meerut RRTS and additional RRTS (phase one) corridors | 209 |
| 17.2 | Tuk tuk in Delhi | 211 |
| 17.3 | New Delhi railway station | 211 |
| 17.4 | The Delhi-Meerut RRTS | 212 |
| 17.5 | The RRTS station waiting environment | 212 |
| 17.6 | The low-density city of Delhi | 213 |
| 17.7 | Station area redevelopment opportunities | 213 |
| 18.1 | Shenzhen's Metro system | 216 |
| 18.2 | High-density city development | 219 |
| 18.3 | Mass transit and an EV bus fleet | 219 |
| 18.4 | A pilot city for the EV bus fleet implementation | 220 |
| 18.5 | EVs in the private vehicle market | 220 |
| 19.1 | Metrolink and new urban development in Manchester | 227 |
| 19.2 | Deansgate station and surrounding development | 233 |
| 19.3 | Rochdale Canal and Beetham Tower | 234 |
| 19.4 | Castlefield Basin and Merchant's Bridge | 234 |
| 19.5 | Metrolink in the city centre | 235 |
| 19.6 | Salford Quays and Media City | 235 |
| 19.7 | Ancoats and New Islington | 236 |
| 19.8 | Metrolink in Droylsden | 236 |
| 19.9 | Deansgate and cycling provision | 237 |
| 19.10 | Manchester Piccadilly railway station | 237 |
| 20.1 | The Valenciennes tramway | 240 |
| 20.2 | Valenciennes station tramway stop | 241 |

| | | |
|---|---|---|
| 20.3 | The tram passing in front of the railway station | 242 |
| 20.4 | Landscaping to reduce the visual impact of the tram | 242 |
| 20.5 | The single-track bi-directional system minimises land-take in the urban area | 243 |
| 20.6 | Neighbouring towns are made accessible by the tramway | 243 |
| 20.7 | The university is also served by the tramway | 244 |
| 21.1 | New urban spaces along the Medellín River | 247 |
| 21.2 | Cable car and integrated Metro and BRT in Medellín | 249 |
| 21.3 | Overground Metro | 250 |
| 21.4 | The BRT system serves suburban neighbourhood centres | 251 |
| 21.5 | Parques Del Río waterfront renewal | 251 |
| 21.6 | Public spaces for pedestrian use | 252 |
| 21.7 | The Encicla system | 252 |
| 21.8 | MetroCable serving the hillside informal neighbourhoods | 253 |
| 21.9 | Escalator access to and from Comuna 13 | 253 |
| 22.1 | LTN21 West Ealing-South | 258 |
| 22.2 | Wooden planters to block through traffic movements | 264 |
| 22.3 | Improved walking and cycling environments | 264 |
| 22.4 | Opposition to LTN21 | 265 |
| 22.5 | LTN21 is removed and the car returns to dominate the street | 266 |
| 22.6 | Richmond Park and cycling during the Covid-19 lockdown | 267 |
| 23.1 | Portland and highway infrastructure provision | 270 |
| 23.2 | Portland overlooking the Willamette River | 279 |
| 23.3 | The light rail system, the Metropolitan Area Express (MAX) | 280 |
| 23.4 | Cycle network in the city centre | 281 |
| 23.5 | Waterfront Park | 282 |
| 23.6 | Redevelopment at the Yards at Union Station | 283 |
| 23.7 | Highway infrastructure | 283 |
| 23.8 | The I-5 interstate highway | 284 |
| 23.9 | Historic housing in Albina | 285 |
| 23.10 | Freeway access direct into the city centre | 286 |
| 23.11 | Severance problems associated with I-5 interstate highway provision | 287 |
| 24.1 | A normative and participatory transport-planning process | 314 |

# List of tables

| | | |
|---|---|---|
| 2.1 | Discourse and international case studies | 39 |
| 24.1 | A discursive framework for understanding transport systems | 295 |
| 24.2 | Dimensions of discourse by case study | 302 |

# Foreword

Much has been written on transport planning over the last one hundred years, from the founding fathers of urban planning in the 1920s to the current concerns over the role of the car in the city. Originally, the car was seen as a means of liberation, for those who could afford to drive, and it quickly became a symbol of the optimism and affluence of postwar Europe. It had already achieved that iconic position in North America, and the same is now happening in China, India and South America. The car in its many forms has become the dominating feature of many cities. Since mass motorisation, there has never been any concerted effort to limit its role in cities as the motor industry and the rich have always found ways to promote car sales to an ever-increasing market, and when one market became saturated, the solution was to expand the existing clientele and to exploit new markets overseas. This business model has lasted for decades, as the car industry has always enjoyed powerful political support. Moreover, the industry has provided employment, profit and tax revenues, and it has benefited from the strong support of the oil industry and more recently technologists through the ever-increasing demand for fuels – diesel, petrol and most recently electricity.

Even when the many disadvantages of high levels of car dependence become apparent, this has done little to dent the dominant narrative. The environmental and social costs have been well documented, and even the diminishing efficiency of the car has not impacted sales. The most recent debates over the impacts of cars on cities and the concepts of sustainable urban mobility have merely acted alongside increased reliance on the larger and heavier electric vehicle. The impacts of motorisation on cities are manifest and contribute to reducing the quality of urban life. They include emissions (carbon dioxide), air quality (pollution, noise and health impacts), open and green spaces (parking requirements), safety (crashes), congestion (allocation of streetspace), well-being (sense of isolation), affordability and alternatives (public transport, walking and cycling), and the ability to drive (about 30 per cent of people cannot drive) – all these have been acknowledged, but with little real universal action or even a willingness to change. There is a huge inertia within the current system, with strong advocacy from the industry and pressure on government to only make marginal changes to address these wider environmental and social costs.

These issues and many others have motivated Rob Hickman to draw on his extensive practical experience and more recent academic expertise to present a powerful case for a fundamental reassessment of transport planning and for the adoption of an alternative paradigm. His starting point is a rejection of the current transport-planning approach based on simplistic quantitative analysis that seeks to reduce city reality to solutions that reinforce car dominance and narrowly defined economic benefits. The alternative proposed is to promote a richer, more diverse perspective that leans heavily on Foucauldian discourse analysis, which focuses on power relations in society as expressed through practices and language. The book is structured around Foucault's six elements – history, truth, power, discontinuity, ethics and subjectivity. After introducing the reader to the concepts, each element is illustrated with evidence from 20 case study cities where new thinking on sustainable mobility has been introduced in different ways, depending on local circumstances and priorities. The message here is essentially positive as there are many examples of good news from around the world, as illustrated in the central city-based chapters of the book.

As noted previously, the traditional approach to transport planning, based on economic efficiency, rationality and political neutrality, is rejected in favour of a more normative and participatory approach embedded in discourse analysis, and extensive and continuous engagement of different groups of actors. This engagement is seen as an essential element to establish the context and vision for your city, discuss strategic options, and finally address appraisal and the broader social and environmental issues.

Such a structure addresses the concerns over the well-established technocratic structure of transport planning, and it outlines a pathway that is specific on context, realistic on objectives and inclusive in its application. Rob speaks of his frustration with the existing and accepted approaches to transport planning, and he argues passionately that it can be viewed as a misuse of power and knowledge, as it only serves to perpetuate the motorisation agenda. He has struggled over the last 20 years to find an alternative theoretical paradigm. Discourse analysis takes the debate beyond critique to one that outlines 'enlightened' pathways to the achievement of broader public policy objectives, and one that places sustainable urban mobility at the centre of collective decision making.

This book provides an accessible and informative text that combines academic rigour with excellent detailed city descriptions, quotes and conclusions. There are also many maps, illustrations and figures to help the reader navigate the complexities of the narrative and the main messages of the book. This means that its appeal is wide as it is addressing many different audiences. At the end of the book, Rob asks two questions: What would a city and its region look like without the car being used as the main means of travel? What might life be like if most of our trips were made using public transport, walking and cycling? Answering these questions would give us an idea of what the transition to sustainable urban mobility might involve. Although Rob cannot give us the definitive answers, he does provide the framework within which to address the alternative visions and strategic objectives, the options available, the implications for accessibility, fairness and the quality of life, the economic, social and environmental costs, and issues relating to implementation, appraisal and monitoring. Read the book to find out more!

*David Banister*
Emeritus Professor of Transport Studies
School of Geography and the Environment
University of Oxford

# Preface

In 2003, I visited Freiburg for the first time. I was working for a well-known and enlightened architecture, planning and urban design consultancy, Llewelyn Davies (LD), now sadly swallowed up by one of the major planning consultancies. LD was of that rare consultancy type: interested in understanding and promoting the quality of urban planning and design, rather than focusing on making money out of the projects worked on and the planning system worked in. Martin Crookston, Tim Pharoah, Patrick Clarke, others from LD, some private developers, contacts from the ODPM (the Office for the Deputy Prime Minster – the then national planning department) and I, travelled over to see the innovative new neighbourhoods being built in Vauban and Rieselfeld. Some had seen them before, but others not, including me. We were all quite amazed to see the new developments, connected to the city centre by tram and segregated cycle networks, and to other cities by high-speed rail. The neighbourhoods were very different to anything I had seen previously, with very low levels of car access and usage, reduced car parking provision, and extensive community spaces in between the housing, utilising the spaces not used for cars. The housing was designed with varied architectural styles, some by residents and architects as self-build; there were affordable housing and community facilities, all set within beautiful landscaped, open spaces. We all asked the following questions: How could they do this here? What were the motivations and how did the city authorities gain agreement with developers, wider key actors and the public? What is it like to live here? And, why could we not do this in the UK?

Often, we take our MSc Transport & City Planning students to Freiburg on the annual course field trip, and they are usually similarly impressed. This revelatory field trip has been repeated in different cities over the last 20 years and always I ask the following: How was this possible, how did the local transport-planning and urban-planning practitioners develop their innovative ideas, what was their inspiration? What was special about the context that facilitated implementation? Could some of the practice be applied elsewhere? It is these types of questions that this book focuses on – examining the very varied practices taken by different cities in transport and city planning.

There are many interesting dimensions, including that most cities remain largely car-dependent, and, strangely, that this remains mostly unquestioned by politicians and the public. It seems that we are not aware of the great possibilities on offer for our transport systems and cities, and we are unable to develop or request better ways of travelling and living. The cities examined in this book arise from very varied political and cultural contexts, and they provide different narratives for transport and city planning. They challenge the discourse of motorisation and provide alternative discourses for sustainable urban mobility.

# Acknowledgements

Thanks, as ever, to Helen, Martha and Oscar, for allowing me to retire to the loft, to listen to some music and do some writing.

Huge thanks to all those who took me around the various cities and very generously gave their time to explain how the transport systems work and how the various projects and urban developments had been planned and implemented; and others who gave their considered views on particular cities:

Plymouth: *Jon Shaw*
Oxford: *Peter Headicar, Martin Kraftl, Craig Rossington*
Freiburg: *Wulf Daseking, Iqbal Hamiduddin*
Singapore: *Paul Barter, Timothy Toh, Li Bin Toh, Chris Donaldson*
Bogotá: *Darío Hidalgo, David Uniman, Thomas van Laake, Natalia Tinjacá Mora*
Houten: *André Botermans, Kylie van Dam, Robert Derks*
Chongqing: *Lixun Liu, Matthew Cao, Ting Peng, Cong Chen, Yantao Ling, Qingzhong Ren, Weidong Song*
London, King's Cross: *Peter Bishop*
Rio de Janeiro: *Clarisse Linke*
Utrecht: *Mark Wagenbuur, André Pettinga*
Copenhagen: *Andreas Røhl*
Malmö: *Ewa Westermark*
Dar es Salaam: *Mohamed Kuganda, Chris Kost, Ronald Lwakataware*
Delhi: *Sharad Saxena, Vinay Kumar Singh, Sidwin Hegde, Samir Kumar Sharma, Mukut Sharma*
Shenzhen: *Joe Ma, Jiawen Yang, Leonard Chew*
Manchester: *Richard Knowles, Ben Brisbourne, Mia Crowther*
Valenciennes: *Philippe Roulet, Adrien Ferrandez, Nicolas Lucas*
Medellín: *Lina López Montoya, Jorge Ramos, Jorge Pérez, Daniela Trejo Rojas, Jose Richard Blanco, Juan Pablo Giraldo, Julio Dávila*
London, Ealing LTN21: *Andrey Afonin, John Dales*
Portland: *Jennifer Dill, Aaron Golub, Aaron Brown, Chris Smith, Ben Crowther, Megan Channell, Anna Howe*

Further thanks to Daniel Moser, Verena Flues and Viviane Weinmann, from the Deutsche Gesellschaft für Internationale Zusammenarbeit (GIZ), for commissioning the work on Transforming Urban Mobility, developed as an online course with UCL and Futurelearn, from which some of the material for the case studies was drawn. Thanks to Matt Aucott, Digital Media, UCL, for video filming, production and editing for the online course, and some of the photography used here, and to Jo Stroud, Digital Education, UCL, for pedagogical guidance on the online course. Thanks to Claudio de Magalhães, Mike Raco and the Bartlett School of Planning, UCL, for providing the space, time, encouragement and inspiration to research and write, and for the sabbatical that allowed me to complete the book. Thanks to Chris Penfold, Elliot Beck and an anonymous reviewer at UCL Press for the smooth commissioning and production of the book.

Thanks to Jamie Quinn for mapping and graphics and Duncan Smith for the image on motorisation rates.

## Image credits

All photographs and images are from the author unless otherwise listed in individual figure captions. Thanks to the Advertising Archives for car advertising prints and Bridgeman Images for the art print.

We have made every attempt to agree publication permissions wherever possible, but would be glad to correct these if we have overlooked particular rights.

## Publication credits

The views expressed, and any errors, are from the author and do not necessarily reflect those of any of the organisations or individuals who kindly gave funding, data, inputs and comments. Some of the chapters have seen light at various stages of development as a journal article or book chapters, as listed below. Thanks for permission to various publishers to use some of the material from these. Thanks to the *Town & Country Planning* journal for allowing reuse of some of the ideas initially developed in a series of articles in the column 'Off the Rails'.

Chapter 9, Houten, draws from a lengthier journal article: Hickman, R., Lu, P. and Botermans, A. 2025. 'Cycling and discourse in Houten'. *Journal of Urban Design* 1–21.

Chapter 22, London LTN21, draws from lengthier book chapters: Hickman, R. and Afonin, A. 2022. 'Transport $CO_2$ mitigation and the production of low traffic neighbourhoods: Lessons from London'. In *Climate Change Mitigation: Policies and Lessons for Asia*, edited by D. Azhgaliyeva and D. Rahut, Tokyo: ADBI.

And, Hickman, R. and Afonin, A. 2025. 'Understanding the opposition to low traffic neighbourhoods'. In *Handbook of Transportation and Public Policy*, edited by A. Perl, R. Ray and L. Reardon, Cheltenham: Edward Elgar.

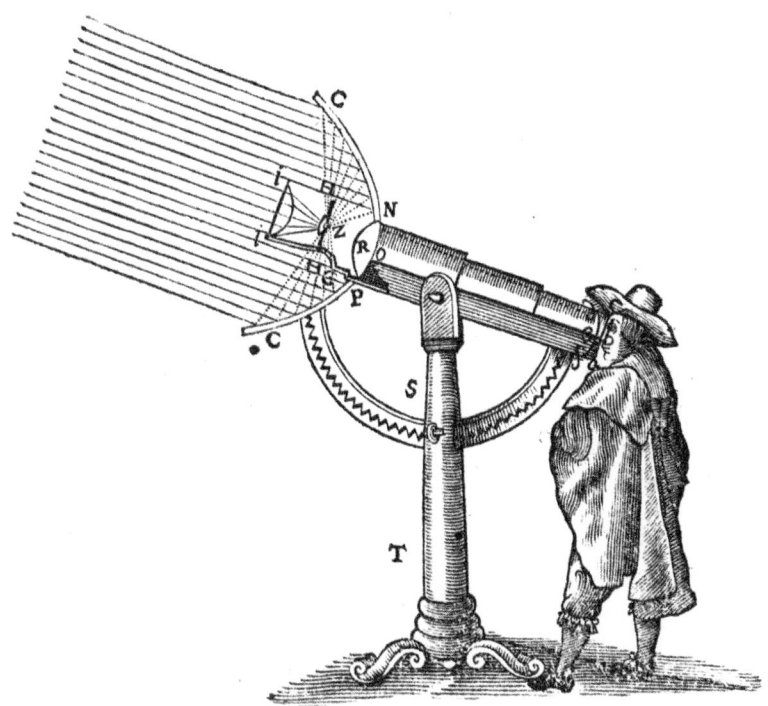

**Figure 0.1** Illustration from 'La Dioptrique' by Rene Descartes (1596–1650). Source: From *Discours de la Methode*, first published in Leiden in 1637 (engraving) (b/w photo), French School (seventeenth century). With permission of Bridgeman Images.

Part I
**Introduction**

# 1
# The production of mobility

> *There is a battle for truth or at least around truth ... by truth, I mean the ensemble of rules according to which the true and the false are separated ... a battle about the status of truth and the economic and political role it plays. (Foucault, 1977, 131)*

## What transport systems are produced and why?

We like to think of society as progressing over time, particularly in the cities where we live. Yet, in many areas of public policy there seem to be many difficulties in transitioning towards new pathways. Transport, for example, has a critical role in contributing to the public policy goals of climate change, social equity and well-being. But, transport systems, across multiple contexts, are not achieving sustainable urban mobility to any significant degree – there is too little progress towards sustainable travel behaviours. Transport systems are remarkably varied across different cities, leading to very disparate travel behaviours at the individual, neighbourhood and urban levels. There is sometimes extensive usage of public transport, cycling, walking or informal transport in specific contexts. But, mostly, cities are designed around highway networks and there is dominant usage of the private car, particularly in suburban neighbourhoods. Once cities are designed for motorisation, the resulting high levels of vehicle kilometres are difficult to progress beyond. There is even evidence of 'forced' car usage, as people spend significant proportions of household income on car travel due to the lack of alternative means of travel. Cities, hence, take different developmental pathways for transport. What is designated as a reasonable transport system in one context may differ in another, and the production of transport systems and travel behaviours becomes a social construction (Berger and Luckman, 1966).

A social construct can be seen as existing not only in objective reality, but also as a result of human interaction – it is the meaning given to an intervention that shapes the intervention. Dominant groups and actors influence what is deemed to be 'true' and normalised for transport planning in particular contexts. The construction of knowledge and reason also determines what is unreasonable. These value judgements, made by transport and urban planners, politicians and the public(s)[1], reflect what is selected and produced as transport systems, but also what is not selected and not produced. Therefore, transport and urban planning can be viewed, in these terms, as a 'social construct': the transport system is shaped by and helps shape society. Alongside, the technical procedures of transport and urban planning help guide what is possible to deliver. But, this is undertaken within the historical and cultural context framing project development.

Proposed new transport projects and urban developments are sometimes characterised by conflict between different groups who do not share the same values or aspirations for public policy. Different groups have uneven possession of knowledge, resources, power and status, hence transport planning becomes much more than a technical exercise, and instead reflects political decisions made over decades. Often there is little consensus over the normative propositions that guide policy planning and the decisions made remain untransparent. This is the central problematic of this book: to understand why transport systems and travel behaviours are shaped as they are, including the narratives given and the impacts that these have. In practice, transport planning is much more than a question of how agreed objectives can be most effectively delivered; it is more than an objectives-led process. The trajectory of transport planning, including the strategies and projects that are implemented, reflects hidden social conflicts. Hence, transport planning is firstly a problem of why some objectives, rather than others, are and should be sought. It is this critical debate that can be made much more evident as strategies and projects are developed.

The historical planning and implementation of transport systems, and the relative provision of different modes, is seldom the subject of debate by city populations. There may be some marginal concern on issues, such as the levels of congestion, or perhaps late-running buses. But, there is little fundamental, critical questioning of what transport networks are being provided, including the extent and quality of transport provision, the shape of the built environment, and how this relates to travel behaviours, experiences and activity participation. There is little debate on the environmental and social impacts of the transport systems provided and to what extent environmental or social goals are being achieved. There is little awareness of the significant improvements to transport systems that might be possible, including the examples available in different contexts. This is despite the varied impacts that different transport systems have on climate change, social equity and well-being, indeed, more generally, on people's qualities of life. The public tend to adapt to whatever travel systems are produced and live their lives without much fundamental questioning of the approaches taken. Sometimes, there are protests against new projects put forward, even where they may be positive, such as for environmental or social goals.

History, in transport planning, is not a record of events, reflecting the delivery of successful projects. It is more a question of how transport has been provided, and why, including who and what has been included and excluded. It is the role of the transport and city planner to be much more aware of these processes, to be conscious of the potential transport systems available and the public policy goals that need to be achieved, in other words, to utilise critical analysis. Much more innovative strategies are required for travel behaviours to be consistent with environmental, social and economic goals. Yet, there is little discussion concerning the serious public policy challenges for society, for example, contributions towards resolving climate change or improving social equity. We sometimes see quite vociferous campaigns against the reallocation of streetspace away from private cars, yet little concerning climate change and the limited role that transport is playing in reducing carbon dioxide ($CO_2$) emissions. This seems the wrong balance in debate on transport planning. It should, therefore, be the role of transport and city planners to make these contradictions more transparent and explicit, so that there is wider awareness and critical discussion on transport systems, travel behaviours and their impacts. This can be part of a process of developing future visions for travel behaviours, with more effective public participation, to ensure there is greater discussion and consensus on the problems faced and the pathways forward. Much of this can involve stronger participatory and deliberative approaches, so that the public are aware of the great public policy challenges being faced and of the possibilities for transport systems and individual behaviours to contribute much more strongly.

The shape of transport systems reflects the social and political culture found in specific contexts and times. The modes available, the journeys possible, the travel experiences and the activity participation that follow, can be viewed as the narrative of the transport system within the particular cultural context. We are beginning to understand that travel is much more than the journey from A to B. The transport system is the means to activity participation, and there are also the impacts of the journey, including the experiences of the journey and the environmental, social and economic impacts of transport and travel behaviours. Drawing on authors such as Bourdieu (1991), on the use of language in relation to power, and applying these concepts to the transport domain, we can understand that transport provides linkages between activities and is a contributory factor in social interaction and communication. Transport can be viewed as a medium of power through which individuals and organisations pursue their own interests. This element of power in transport planning is a critical, yet rarely considered, element of the transport systems produced. Foucault (1969) was interested in power and knowledge, and how these are used as forms of social management and control. These issues can be applied to transport planning and used to broaden the theoretical understanding of transport, beyond the technical exercise of planning and implementing transport systems and projects. In this way, we can seek to understand what transport systems mean for travel behaviours, activity participation and the social order that this represents and shapes. As part of this, there is an inherent problem of the reliance on rational systems of thought within transport planning. Through these perspectives, critical views on transport systems are lost and dimensions of power and knowledge are overlooked. As Foucault (1982a, 358) explains 'What is this reason that we use? What are its historical effects? What are its limits and what are its dangers?'

The central paradox for transport planning is to balance the perceived aspirations for individual mobility with the well-being of society. In almost all contexts, the tensions remain unresolved: there is a major implementation gap between public policy on sustainable urban mobility and what transport systems are achieving in terms of travel behaviours and impacts (Hickman and Banister, 2014). Consider climate change and the need to reduce transport $CO_2$ emissions: transport is the key sector making little contribution to reduced $CO_2$ emissions. Even in the so-called 'progressive' transport cities, transport $CO_2$ emissions are decreasing only marginally. In almost all other contexts, transport $CO_2$ emissions are rising, sometimes dramatically. Consider traffic casualties and how there are many vehicle-related deaths and casualties in every city. Somehow this seems to be overlooked and the casualties have been 'normalised' as people carry out their everyday lives. Consider levels of inactivity in relation to transport, with rising obesity, particularly in younger age groups, and subsequent rises in non-communicable diseases (NCDs). Consider how transport systems are used by some, but not others, allowing or not allowing travel to specific parts of the city and participation in particular activities. Consider how urban planning relates to transport and often facilitates specific forms of travel, such as lower income groups being forced to the edge of the city, or beyond, leading to lengthier journeys and times, or suppressed trips relative to aspirations. The social equity dimensions of travel are very significant, with activity participation remaining difficult or impossible for some. The use of transport is, hence, very unevenly distributed spatially and across population groups. These impacts from transport, often seen as 'secondary' outputs, are usually overlooked by transport planners, as the analytical focus remains on assessing realised travel behaviours, and frequently at the aggregate level. Is the transport system successful if it is only used by particular groups, overlooking those who cannot use it and why? Is an accurate estimation of demand and cost important, irrespective of wider impacts? Well, largely, this is how transport systems are currently judged.

I give you some key statistics that, each by themselves, should make us reconsider our approaches to transport planning – and to move more effectively towards sustainable urban mobility systems:

- Energy consumption: the transport sector consumed around 2,200 million tons of oil equivalent (mtoe) in 2015, equating to 23 per cent of energy usage, growing at nearly 2 per cent per year over the last decade (International Energy Agency, 2016). Much of this is from finite, oil-based energy supplies.
- Climate change: in most countries, even where overall $CO_2$ emissions are decreasing, the level of transport $CO_2$ emissions is rising. Global transport $CO_2$ emissions have risen by 65 per cent from 1990 to 2021, and even increased by 16 per cent in the EU-27 where the growth in $CO_2$ emissions has plateaued (Crippa et al., 2022).
- Traffic casualties: over 1.2 million people die globally each year as a result of road traffic crashes and road traffic injuries (World Health Organization, 2023). Traffic casualties are the leading cause of death among young people aged 5–29.
- Inactivity and health: NCDs killed around 38 million people in 2012, representing 68 per cent of 56 million global deaths (World Health Organization, 2015). Inactivity and levels of obesity contribute to NCDs, and a significant part of this is related to private car-based travel and dispersed built environments.

Add in the adverse impact of motorisation on streets and across neighbourhoods and cities, with too much space given to the car, and little for other modes, and we have some serious problems to resolve. Yet, the discussion of transport in the popular media and amongst politicians and the public is often frustratingly pro-motorisation. There is a framing of narratives that result in many people understanding that streets should 'naturally' be used for the use of private cars and that more traffic capacity is required. This shapes the general knowledge (the *savoir*) on transport and travel behaviours; and many politicians and a significant proportion of the public support the use of the car for their everyday activities. Indeed, many transport planners work to improve the highway network in their daily jobs, as this is where the highest proportion of consultancy funding is made available. Many transport planners are taught and seek to analyse and optimise motorised travel behaviours.

What is happening here? How can the practice of transport planning be so contradictory to public policy goals? Why is there little problematisation of the adverse impacts? This is viewed as being deliberately maintained by the existing apparatus and actors to facilitate their activities (Culver, 2018). Transport planning, as a discipline, is focused on improving mobility largely through motorisation, alongside the provision of insignificant investments in public transport, walking and cycling, across multiple contexts. Look at almost all cities globally and most have made similar mistakes of motorisation, albeit at different stages in their histories. Many cities have provided too much space for the private car, at the expense of other users of the street. Many do not invest sufficiently in public transport or walking and cycling, and urban planning is not utilised to support public transport and active means of travel. This is the scale of the challenge: transport planning has to contribute more significantly to reducing $CO_2$ emissions, to facilitate more equitable travel and access to activities, to improve health through reducing traffic-related casualties, and to improve well-being through active travel and activity participation.

Not all amongst the population are willing to support more sustainable transport systems and behaviours, as the system of motorisation is deeply embedded in many people's lives. Think of your ubiquitous local traffic reports, the coverage that is given to traffic congestion, and the impact that this might have. Take an example from London, where popular sources for live travel status updates include MyLondon and Transport for London (TfL). They give us the following news feeds on one particular morning (Wednesday 15 March 2023, 9am):

- [A13] crash live: Crash with lorry and motorbike blocks major east London road causing huge five-mile queues in rush hour.

- [M23] crash live: Crash near Gatwick Airport sparks five miles of queues.
- [A5] Edgware Road traffic live: Emergency vehicle incident shuts major West London road.
- [M25] traffic live: Two lanes blocked and severe delays due to accident, two lorries involved on M25 anticlockwise between J26 A121 (Waltham Abbey) and J25 A10 (Enfield). Congestion to J27 M11, and heavy through Waltham Abbey. Travel time is 1 hour 20 minutes.
- [A13] Commercial Road (both directions) at the junction of Butcher Row: serious delays.
- [A40] Western Avenue (both directions) between Polish War Memorial and [A312] Target Roundabout: serious delays.
- [A406] North Circular Road (Westbound) at the junction of [B453] Neasden Lane: serious delays.
- [A102] Blackwall Tunnel traffic live: minor delays.
- [A10] Kingsland High Street: minor delays.
- [A4] Piccadilly Underpass (W1J/W2) (both directions): underpass has lane reductions in place to facilitate maintenance works: delays are possible.
- [A306] Hammersmith Bridge (W6) (both directions): bridge closed to all modes of transport apart from pedestrians and cyclists.
- Drivers warned [A102] Blackwall Tunnel is closing for two days for repairs (10pm on Friday 31 March until 5am on Monday 3 April) with motorists told to use Dartford Crossing instead.

And, as the following extra news articles:

- London ULEZ: Drivers could face 'up to £3,250 worth of fines' after £12.50 charge is extended.
- London ULEZ: Vast majority of Londoners have absolutely no intention of replacing old cars to avoid £12.50 daily charge.
- London ULEZ: Grandad's heartbreak at being 'unable to visit' seriously ill granddaughter, six, 'living on borrowed time' as he can't afford £12.50 'tax'.

A strong narrative is produced, from supposedly 'objective' and even 'authoritative' news sources, which reinforces problematic travel behaviours. The messaging from MyLondon is almost completely concerned with traffic congestion, giving the impression of a 'crisis' on the highways. Even the updates from TfL, which has a central remit for achieving sustainable transport, usually focuses on traffic congestion. Yet, this is simply a metric highlighting the unreliability of the highway network for vehicles, and 'naturally' leads us to consider how we can improve the flow of traffic. This was not a special day for congestion in London, but an impression is given of a highway network requiring much more funding for increased capacity and reliability. The implied requirement for transport planners is to reduce the congested traffic movements by providing more space for traffic. These types of live traffic updates will even break over the music playing in your office or car, highlighting the urgency of the situation. There is no related discussion of the absence of local bus services, the cost of public transport, or the lack of safe cycle or pedestrian networks. Or, the inequity of transport provision and activity participation over space and time. Or, indeed, the likely adverse impacts of more traffic.

The discussion is even extending into populism (Gössling et al., 2024) and the so-called 'culture wars', as political positions are taken against charging for the use of vehicles through the extended Ultra Low Emission Zone (ULEZ). Rishi Sunak, as the previous UK Prime Minister (until 2024), questioned the use of low traffic neighbourhoods (LTNs) in the UK. There was similar discussion on the 15-minute city concept and lower speed limits, drawn together in the UK Department for Transport's (2023) preposterous 'Plan for Drivers'. A number of populist politicians are utilising transport as a 'wedge issue' in an attempt to increase political support. A wedge issue is a topic with a controversial and divisive nature, raised in an attempt to gain votes from the opposition. In transport, the car driver is seen as a possibility for vote winning, with a cohort of people more interested in the convenience of car usage than in environmental issues. In this case, it was the Labour-voting 'forced' car owners with lower incomes, who might be attracted to voting Conservative if more funding was given to improving traffic conditions. The tactic gained little traction in the 2024 General Election in the UK, but demonstrates how the implementation of transport policy has become complex and messy, reflecting the increased social conflict and polarisation associated with public policy. The current Labour government has failed to tackle car dependency through increased fuel taxation or major investments in public transport or cycling infrastructure. The urban and regional planning regime remains weakly applied and resourced. The government seems very reluctant to annoy the car driver. Move over to the USA and witness the tumultuous presidency of Donald Trump. There is a very significant reduction in environmental regulations, support for the oil industry and motor manufacturing, and removal of funding for public transport and cycling projects. We can see that transport and urban planning remains ineffective in the midst of a difficult political and cultural context. Motorisation continues to gain political support and sustainable urban mobility behaviours are unlikely to be achieved. Meanwhile, transport planning remains focused on rational procedures, such as the techniques of project planning and design. The expectation is that this will lead to the implementation of projects, and achieve a significant transition towards sustainability, yet it overlooks the social conflict inherent in strategy and project development.

According to INRIX[2] (2021), London has apparently become the most 'congested' city in the world as drivers 'lost' the most hours globally (148 hours); worse than Paris (140 hours), Brussels (134 hours), Moscow (108 hours) and New York (102 hours). In the UK, drivers apparently lost a total of 73 hours each in travel, up from 36 hours in 2020. Traffic levels have returned to pre-Covid-19 levels, as people return to commuting, often facilitated by the car. The resulting cost of traffic congestion, as estimated, is around £595 per driver per year in 2021, up from £303 in 2020. In aggregate, this is assumed to cost the UK £8 billion per year. This type of metric is widely used across different contexts and illustrates the focus given to supporting motorisation. Measuring and alleviating congestion (usually defined as the time spent on a journey in congested conditions relative to free-flowing conditions) means that the case is made for freer-flowing traffic. The emphasis on individual and aggregated travel time and cost, and the assumed direct linkage to economic growth, is rarely questioned. The assumed 'economic cost' of congestion and linkage to the economic success of cities is, at best, poorly evidenced, with little discussion of how travel time, if realised, is used (Metz, 2008). This seems a fundamental flaw in the understanding, yet it does not lead to a revisit of the logic. The solution put forward is that highways should be improved and traffic capacity increased, so that individuals can experience speedier travel conditions. Subsequently, their lives will be improved. Hence, the congestion metric is chosen to provide evidence for a particular policy stance. There are few related discussions on the inadequate state of public transport, cycling or walking provision. In many contexts, it is impossible or implausible to travel by public transport, walking or cycling, yet this remains unmeasured and undiscussed. The wider adverse impacts of motorisation and inadequate public transport, walking and cycling, including multiple environmental and social costs, remain overlooked. The relationships between the extent of these transport networks and economic and quality of life goals remain unknown. Congestion is simply a metric created to help justify improvements to the highway network. But, it remains central to the vast majority of transport strategies, despite little reflection that a normative position is being given by this form of analysis. There are related and wider issues surrounding knowledge in transport planning, including through the technical procedures and approaches used. These also need to be more critically discussed and reconsidered.

The practice of transport planning, therefore, becomes a complex problem of governance and mediation between viewpoints, moving beyond the conventional technical analysis of travel behaviours. This requires a broader consideration of the most effective planning approaches and implementation of transport projects, consistent with public policy goals. Foucault (1991, 93), drawing from Guillaume de la Perriere, considers governance as producing 'the right disposition of things arranged so as to lead to a convenient end'. Transport planning can learn much from this type of thinking, considering what the right disposition of transport might be, relative to the convenient ends that are required. Think of a visit to a city with a high-quality public transport system, or excellent walking or cycling facilities. Perhaps you have visited cities in the Netherlands or Germany. How have these transport systems and subsequent travel behaviours been produced? What are the impacts of the travel behaviours and the participation in activities that follow? Why are transport systems and travel behaviours so different in these contexts? Most would agree that the vast majority of urban areas in the UK do not have effective public transport, walking and cycling networks, yet there is little debate concerning how this can be changed.

## Focus of the book

This book focuses on these issues: to understand why transport systems and travel behaviours are shaped as they are, differing in space and time, including what is deemed as right and wrong, or appropriate, in particular contexts. Many cities are decades behind in the transition to sustainable urban mobility and there seems little urgency to change. We should much more critically question the narrative, as developed, or undeveloped, by the city authorities, the politicians and the public, considering how and why particular contexts support motorisation and limited investment in public transport and cycling, or produce differing forms of more sustainable urban mobility.

The commentary challenges the dominant discourses encouraging the processes of motorisation and the ineffective implementation of sustainable mobility options. This is done by examining how transport systems have been produced in different city case studies. The narrative starts in Plymouth, to examine a classic case of motorisation, with its comprehensive rebuilding of the city's urban fabric around a revised highway system from the 1940s onwards, after the bombing of World War II. Further cities, following different trajectories, are then examined, including Oxford, Freiburg, Singapore, Bogotá, Houten, Chongqing, London (King's Cross), Rio de Janeiro, Utrecht, Copenhagen, Malmö, Dar es Salaam, Shenzhen, Manchester, Valenciennes, Medellín, Delhi, London (West Ealing-South) and Portland, each of them offering different pathways forward for transport and city planning.

The book aims to

- Bring discourse analysis into the field of transport and city planning, developing a framework and discussing a range of discursive concepts, such as history, discontinuity, power, truth, ethics and subjectivity, which can help to understand the positions taken and the social conflict in transport.
- Examine transport systems and projects in different case study cities, including examples of motorisation in relation to forms of public transport, cycling, walking, urban planning, travel demand management and electric vehicles. The discourses are presented in relation to the wider historical, sociocultural and political contexts.
- Strengthen the case for problematisation, critical analysis and debate around current and potential transport systems, as part of vision-led, participatory and deliberative processes of transport strategy and project development. This will help facilitate greater progress against public policy goals, including climate change, social equity and well-being, and, ultimately, higher qualities of life in cities.

The transport systems in the case studies are examined as discourses, to demonstrate the wider social contexts to transport and to make the case that different pathways are possible in different contexts. The central argument is for much more significant investment in sustainable mobility measures, including urban planning, walking, cycling, public transport and traffic demand management. But, this will require quite different ways of thinking about transport, including understanding the discourses discernible from multiple actors, such as transport and urban planners, engineers, politicians, civil society and the public(s). This provides the start for discontinuity and change in transport systems, applied through a deliberative process of project development. This approach – of bringing discourse into transport planning – offers much hope for achieving more widespread sustainable travel behaviours globally. Through this, we are more likely to resolve the great public policy challenges we face.

## Notes

1. 'Publics' is used to represent the plurality of public views – individuals often have differing views on their local transport systems and varied travel behaviours. Yet the debate on transport systems is usually quite bounded, as if there is only one type of person, and the diverse range of possibilities for different transport systems is rarely explored.
2. INRIX describes itself as 'a world leader in transportation analytics and connected car services'.

# 2
# Understanding space and mobility

## Social order, cities and transport systems

The design of transport systems dominated by use of the private car in cities and across regions leads to very individualised transport behaviours and extensive adverse impacts. The motorised society has become the 'norm' in many contexts, reflecting the transport systems that have been produced by the city authorities. Bourdieu's (1972) view of the habitus helps us understand this: that social and economic conditions and wider cultural norms help shape viewpoints and activities. In this case, they shape the practice of transport planning and the resulting travel behaviours. The predisposition and order of society is hence important to travel. Using the private car as the primary means of travel is not a 'natural' idea, a fact that has to be acknowledged. It has simply become 'naturalised', as an often-prioritised system of infrastructure, and a discourse with wide acceptance, not least by the more powerful actors. There is some critical debate, such as from parts of academia and environmental groups, but this is limited, particularly in view of the scale and impacts of motorisation. Some cities are building extensive public transport systems and cycle networks, and even removing highway provision, but this is far from widespread, and projects are often difficult to deliver. The general point is that increasing highway capacity and motorisation have been accepted as the 'usual' way to build transport systems in cities. Yet, motorisation is simply the social construction produced by the motor manufacturers, oil suppliers, suburban real estate industry and related organisations, as part of the motor industry complex (Freund and Martin, 1993, Paterson, 2007): that urban areas, surrounding regions and travel behaviours should all be designed around the use of the private car. There is a wider set of complicit organisations, including national and city government authorities, highway building agencies, transport consultancies, urban planning consultancies, developers, and even some university transport research departments and wider research institutes. All of these contribute, at least in part, to the motorisation effort. They plan, model, design and implement increased highway capacity and have common financial interests in facilitating increased sales of the private car. Alongside, the techniques of transport planning have been refined and adopted in practice, covering highway modelling, project appraisal and a wider range of approaches, which help build the highway networks. Even in cities with some or even extensive public transport, walking and cycling facilities, the discourse of motorisation is still powerful and limits the investment and space given to other modes. In most cities, car usage is dominant in travel behaviours, for example, in terms of vehicle kilometres travelled by mode. In the more motorised cities, the motorisation narrative is seldom even questioned. The discourse of motorisation, therefore, has had remarkable success in creating the conditions to shape cities and streets, and to sell vehicles. The central role of private cars (and wider vehicles) as expensive products to be consumed, leading to capital accumulation for the key actors in the motor industry complex, is the reason for this dominance of motorisation in transport systems. Incredibly, this is a global narrative, an example of Bourdieu's *doxa* of shared and unquestioned belief (Deer, 2012).

Prioritising motorisation has been the goal of major, powerful actors in the wider capitalist effort. The basic premise was simple: if the population could be persuaded that cities and lives should be designed around the car, then the selling of vehicles could be secured for decades. Convince the public that they should aspire to the latest model of vehicle, ideally the larger, higher specification and more expensive models, expand the market geographically across contexts internationally, and the market would increase year on year. Perhaps this was not an intentional, transparent calculation by only a few actors, but a momentum that grew over decades, serving many vested interests. This motorisation effort has also led to a remarkable consumption of space in many cities, so successful that the use of streetspace for vehicles or space for car parking is rarely critically discussed.

The history of motorisation is well-known, originating from 1886, when Carl Benz patented his motorwagen in Germany. The motor car was then taken to the mass market in the USA by Henry Ford, from the early 1900s onwards. Motorisation has since spread throughout almost all cities and urban areas globally. It has been applied, initially with support from city authorities to convert existing streets for use of the private car and to add new capacity where it was perceived that this was required.

Hence, city authorities and states have been captured in supporting the capitalist effort. In many contexts, the use of streetspace by the car is maintained as city authorities give, or are given, little resource to build effective public transport, walking and cycling networks. There is some contestation over projects, including over the reallocation of space away from dominant car usage, yet the car remains the most convenient option for travel for many people. Motorisation has been successful to the perverse extent of national governments supporting failing motor manufacturers, believing that the motor industry is central to economic growth. The narratives used to support motorisation are wide-ranging and widely accepted, but often implausible. For example, the car is associated with ideals such as 'efficient', 'convenient', 'cheap', 'stylish', 'modern', 'progressive', 'a liberator', 'individualistic' and even 'democratic' – even though there is little substance to these associations, particularly when many people use the car and large traffic volumes result. The motor industry is positioned as critical to economic growth and political success, even though public transport and cycling could equally provide employment and opportunities for technology development and exports. The car has come to dominate everyday travel and urban environments, with motorisation coming of age at varying times in different contexts, from the early 1900s onwards in the USA, and then in many further cities and regions. Forms of urban development were shaped to reflect the functional needs of vehicles, the car becoming the primary means of individual-based autonomy. This has been a great neoliberal success story: our streets, cities and travel behaviours have been shaped around the purchase of an expensive material good, delivered by an array of private operators, including the motor industry and related actors. Alongside, the cultural symbolism of the car has been gradually shaped, with the narrative supported in advertising, films and print media, associating the car with the chosen ideals (Sheller and Urry, 2000, Sheller, 2004, Smoak, 2007).

Norton (2011) describes the gradual reconstruction of the street for the use of the private car in the USA, a move that was initially resisted by the public. Hence, there is a lesser-known narrative of dispute and struggle, unfortunately a losing discursive struggle, as the powerful actors in the motor lobby provided space in cities for the new product. Pedestrians were initially angry with the intrusion of vehicles and demanded restriction of the space given to the car. The police attempted to regulate the use of the street to improve the efficiency of throughflow, using detailed processes of traffic counting, and spent increasing time managing traffic volumes, as 'traffic police'. From the 1930s onwards, the motor lobby pushed for the 'protection' of the pedestrian on safety grounds, restricting their use of space to pedestrian footways and specified crossing points, hence allowing more vehicles to use the city streets. The shared 'motordom' interests of the motor car manufacturers, urban development and insurance industries were coordinated by the American Automobile Association, national and local chambers of commerce and leading motor manufacturers such as Studebaker. An example of reframing the debate was seen in the coining of the term 'jaywalker' to demonise the pedestrian who attempts to cross the road at unspecified points. The streetcar (tramway) was discouraged and removed from the street. The major motor manufacturers, such as General Motors, even bought up the streetcar companies, managing their decline and dismantling urban public transport. Increasingly, more space was given over to the car, instead of attempting to restrict the demand and volume of cars on the street. The traffic safety problem was reframed as a problem for the pedestrian, rather than one of the impact of the vehicle on the pedestrian. Advertising campaigns aimed to modify pedestrian behaviour to improve safety. Though the car is now frequently associated with concepts such as freedom, to many people in the USA in the 1920s and 1930s, the car and its driver were seen as the problems that deprived others of their freedom. Eventually the traffic engineering and city planning professions emerged and conceived their primary task as retrofitting the physical layout of cities to accommodate the car. The street and transport planning were commodified for the selling and use of the car and cities were redesigned and expanded to facilitate this. An extensive interstate highway system was planned in the USA and implemented for travel between cities, while the cities themselves were rebuilt and expanded around the car.

The discursive formation was of motorisation and the reshaping of cities around use of the car. This was an exercise in power, an exchange between social groups, where the motor manufacturers and associated organisations developed the justification – winning the discursive struggle – to reshape streets and cities to further their financial interests. A similar historical trajectory has since spread across many contexts, from North America, to Europe, South America, Asia and Africa, demonstrating a remarkable belief in the motorisation effort. The success of the narrative is seldom critically discussed in the mainstream public debate, with practitioners, politicians and the public rarely considering the utility of motorisation and dispersed form of cities. The associated adverse impacts of mass motorisation on energy consumption, $CO_2$ emissions and local air quality, traffic casualties and inactive lifestyles are all hugely significant and conveniently overlooked (Hickman et al., 2017b). Move forward nearly a century in time, and we find contestation associated with attempts to remove space dedicated to the private car. Streetspace reallocation projects, including the removal of urban flyovers, modified traffic gyratories, new cycle lanes, pedestrian spaces and low traffic neighbourhoods, are often controversial while they are planned and implemented. Some projects fail as they are hijacked by car-supporting lobby groups or cohorts of the public supporting car usage.

Reduced motorisation is more likely to arise with a more transparent and widespread discussion of its adverse impacts. Many of the environmental and social impacts adversely affect human life and, indeed, provide an existential threat to humanity. The use of the car was 'naturalised' as an inevitable feature of life, an icon of 'modernity' and a necessary part of our daily routines. Yet, the motorisation experience and the poor states of public transport, walking and cycling are increasingly being challenged. Some cities have chosen different pathways, shaping urban areas and lives around public transport, walking and cycling networks. There have been previous watershed moments in the debate, for example in response to the energy crises of the 1970s, or when planned new highways were blocked or traffic casualties were highlighted. Yet, vehicle ownership has continued to rise in many contexts, certainly in countries such as China and India, and indeed globally in aggregate terms. The hegemony of motorisation is difficult to break; reallocating streetspace from the car to other modes, raising road 'taxes', investing in public transport, walking and cycling, and planning urban development are all heavily contested and often involve tortuous implementation processes.

A distinction can be made between motorisation and automobility. Motorisation can be seen as the level of ownership and usage of private cars and vehicles more widely, including sports utility vehicles (SUVs), light goods vehicles (LGVs) and vans, and the transport networks provided to serve these vehicles. Whereas, automobility is the wider system of institutions, organisations, actors, knowledge and wider political and cultural factors, making the use of motor vehicles possible, and often necessary (Urry, 2004, Böhm et al., 2006). These actors and mechanisms reinforce and support each other, including the motor manufacturers and associated organisations. They also develop discursive practices, such as the procedures, rules and regulations to support motorisation, and the domains through which the car and motorisation are marketed. Harvey (1973) warned of the features of capitalism in relation to space and cities; that the shape of cities should be understood in relation to the social and political processes that produce them, and particularly in terms of neoliberalism as the contemporary force that produces cities. For example, Oxford was designed as a city of 'dreaming spires', created in the age of the powerful Church and reflected in the dominant form of the skyline. London or New York, in more recent times, have been redeveloped in the age of financial capitalism, adding the contemporary glass and steel of financial institutions to the existing neoclassical architecture in the central city neighbourhoods. More recently, this has been joined by buy-to-let rental and Airbnb properties. In transport, similar trends are evident. For example, motorisation is the outcome of the free market, where cars and highways are seen to pay their own way (though of course there is much subsidy for highway investment and other features), while public transport, cycling and pedestrian infrastructure is seen as being subsidised by the public sector. Transit-orientated developments (TOD), where former railway sidings or industrial spaces are designated as areas for urban development, are converted into new residential neighbourhoods. Often there is very little affordability in the new housing produced, the new public space even being privately managed, and transport connections and accessibility into the city centre are central features of the new developments. Usually, the infrastructure is funded by the public sector, and the use of space and property value uplift is taken only by the higher-income groups. Again, something very interesting is evident: social and political processes are producing a particular shape of cities and transport systems.

Transport systems, travel behaviours and the surrounding built environments are, therefore, representative of the wider social order and time in which they are developed. The historic centres of cities were often walkable, developed before the arrival of the motor car. Many have endured, been protected, and are still attractive to walk around. The inner suburbs in many European cities were built around the growing public transport systems. London was built with suburban development spreading along the new (private) railways, later to be brought together into a more extensive (public) Underground network. Cities and urban areas in the Netherlands were built around cycle networks, with high levels of cycling in compact neighbourhoods. Conversely, many cities in the USA, planned with motorisation as the primary means of travel, were developed around the highway systems and often suburban neighbourhoods were very dispersed and only reachable by private car. These transport systems reflect the importance given to the motor manufacturers and the private car in economic and cultural terms. This motorisation approach to transport planning became hugely influential across many contexts. This includes many cities where the car dramatically conflicts with the socio-economic context or the scale of population, and where traffic congestion can be experienced at absurd levels. This commentary, of course, is a simplification; cities and transport systems have developed incrementally over time, with different means of transport and networks running alongside each other. The resulting travel behaviours are the product of different layers of transport networks available, but also a wider set of contextual factors. Yet, the essential point is that transport systems and travel behaviours do not simply appear, nor do they follow easy-to-understand rules of rationality. They are the result of myriad decisions made by different actors, including transport planners, urban planners and traffic engineers, but also politicians, motor manufacturers, the oil industry, urban and suburban real estate developers, and the public. All of these seek to further what they perceive as the most appropriate way to develop transport systems, with dimensions of power meaning that some become more dominant than others.

This moves us beyond simplistic understandings of transport planning. Dominant positivist analytical approaches, which use only quantitative approaches to study social phenomena, overlook that individuals are often complex and interpret the same 'objective reality' in different ways. Transport systems and travel behaviours do not always follow simple associations or regular patterns and may be the result of contestation and dimensions of power over time. Transport projects are often designed and implemented by transport modellers and engineers in a search for the most 'efficient' transport systems. Transport modelling has conventionally been based on the four-stage modelling process, using the relative attraction of origins and destinations and deriving aggregate travel behaviours from this. But, this is increasingly understood as overlooking important considerations, such as the attributes of the built environment and different modes and attitudes to travel and activity participation and satisfaction with travel (Mladenovic and Trifunovic, 2014). Positivist science gives us a logic of identity, that $a = a$ and $b = b$, that $a$ is different to $b$. Or, that $a + b = c$. In the case of transport, this may mean that the weight of attraction between origin and destination = travel behaviour. Or, perhaps, that transport provision + urban structure + individual attitudes = travel behaviour. Or, that providing cycle infrastructure leads to cycling. Or we could be more sophisticated in the analysis, and examine the strength of relationships, suggesting that $a + bX = Y$, or variations of this.

However, a greater understanding of content and context may mean there are other factors, not estimated, that seem to have significant influences, such as cultural norms or political contexts. There may be differences in opinion; some people might think that $a$ is $b$ or $b$ is $a$, or $a + b = c$ in one context, or $d$ or $z$ in another context. Or, indeed, that transformation can occur and *a* can become *b* over time. Hence, positivist science can become very messy and partial in understanding. There may be other factors in the development of the accepted discourse on the transport system, including contestation and the meaning given to this. There are important dimensions of discourse to be interrogated, such as trajectories of history, truth, power, discontinuity, ethics and subjectivity. Space is not only an abstract container for our lives, but a reflection of the structures, contexts and actors that helped create it (Shields, 1999). The use of space produces environmental and social impacts, including of dependence and exploitation. As Bourdieu (1972) advises, there is an issue of habitus, which is viewed as the practice of actors (such as individuals, institutions and organisations) within the overriding structure of society.

There is, particularly, little consideration of the social equity impacts of travel, beyond the aggregate behaviours, including who uses the transport systems and who is impacted by the resulting infrastructure and travel. The emphasis on forecasting means that public policy goals are not achieved – there is too much focus on providing for existing trajectories. Critically, transport planning requires more analysis concerning how effectively policy objectives are being achieved and can be better implemented. For example, there is often little consensus on what objectives should be delivered and, certainly, to what degree. Many of these complexities are difficult to model, such as the political and power considerations in project planning and implementation. These social and political considerations are poorly understood in project planning and implementation, and often the objectives for transport planning are poorly considered. As a result, transport planning has provided more mobility, and usually for the consumption of already mobile and higher-income groups. This is a critique of transport planning and the purposes it serves, indeed the approaches it uses and the outputs it produces. The challenge is to thoroughly problematise the current status and reclaim the accepted discourse in favour of public policy goals.

## A framework for sustainable urban mobility

Some progress has been made in applying a visioning and backcasting approach (Figure 2.1) (Hickman and Banister, 2014). Future visions can be developed to represent the transport behaviours that we wish to develop. Implementation strategies are developed using the avoid-shift-improve framework (ASIF), including a range of measures to avoid travel

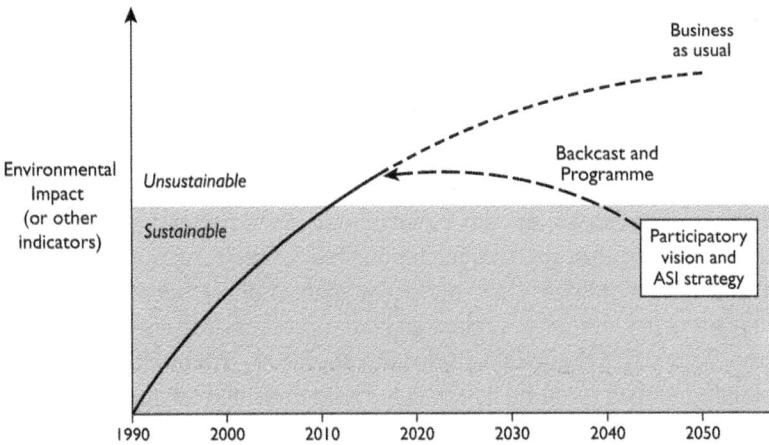

**Figure 2.1** Visioning and backcasting. Source: Hickman and Banister, 2014.

(urban planning, traffic demand management), shift travel (public transport, walking and cycling) and improve travel (low emission vehicles). Pathways for implementation are then developed using a backcasting approach – working back to the present from the future vision. Hence, the forecasting of the 'business as usual' trajectory becomes unimportant, as the extrapolated historic trends are not deemed suitable, even as a reference case. Transport modelling can be revised in focus, to help estimate the future state to be achieved and the most appropriate pathways back from the agreed future.

As part of the development of a revised strategy, there needs to be careful consideration of what public policy objectives should be achieved, including what the best processes might be to meet these. Sustainable urban mobility requires a different set of projects to be planned, with very significant investments required over multiple jurisdictions, consistently over decades. It is significant that 'sustainable transport' remains largely undefined, so that it is unclear as to what is required. The 'fuzzy' nature of public policy goals is visible in many domains, including in relation to environmental sustainability issues, and leads to many different interpretations, including conflicting and contested actions and greenwashing. Sustainability is described as an 'imaginary fantasy' and 'empty signifier' (Swyngedouw, 2007, Swyngedouw and Kaika, 2014), where it is assumed that the economic, environmental and social objectives will somehow be resolved, overlooking the many tensions that are evident. It is helpful to provide a clearer definition for sustainable transport: 'Sustainable transport can be viewed as a means to access activities, within environmental limits and equitably, rather than as a means for a few to consume more mobility' (Hickman and Banister, 2019).

This helps us consider the linkage of travel to activity participation and also the concept of limits in environmental and social terms. But, even here, the contradictions remain unresolved and the environmental limits and thresholds of equity remain undefined for specific local contexts. Without this, there is much potential for differing interpretations and undue emphasis given to particular goals. A clearer set of targets and indicators is required to understand what progress has been made and what progress is required for transport strategies. It is unlikely that sustainable mobility will be achieved without a framework for evaluating progress, leading to reprioritisation of investments. A general aspiration can feasibly be developed into a set of indicators for particular contexts, such as mode share by public transport, walking and cycling, the private car, and other modes; vehicle kilometres travelled, average journey length; energy consumption, $CO_2$ emissions; traffic casualties; built form and accessibility. The contradictions inherent and trade-offs required will hence become more transparent.

Transport strategies and projects should contribute to key public policy objectives such as an improved economy, environment and social equity, as well as well-being and city design (Figure 2.2). Hence, there are societal objectives, beyond the conventional 'three pillars' understanding of sustainability, to help understand improvements to well-being and the wider quality of urban life. The 'nested sustainability' concept can be applied, reflecting that all objectives are reliant on each other and need to be achieved (Giddings et al., 2002, Hickman, 2017). Transport strategies and projects should be developed and prioritised against this broader range and interlinked set of policy objectives. Transport is then more likely to be focused on improving accessibility and activity participation for different groups in society. Transport efficiency, as measured through indicators such as time savings or congestion relief, becomes unimportant, as these are not policy objectives that remain important. We no longer need to measure these impacts, as they are simply the key metrics, amongst others, that have been developed to help build the system of motorisation. Forecasting approaches, as conventionally employed in transport planning, can also be superseded. Instead, discontinuous trajectories are required that can better achieve critical public policy goals.

The difficulties with previous work on scenario analysis and backcasting are that the strategies developed are usually either unambitious or unimplemented and there is insignificant progress made relative to

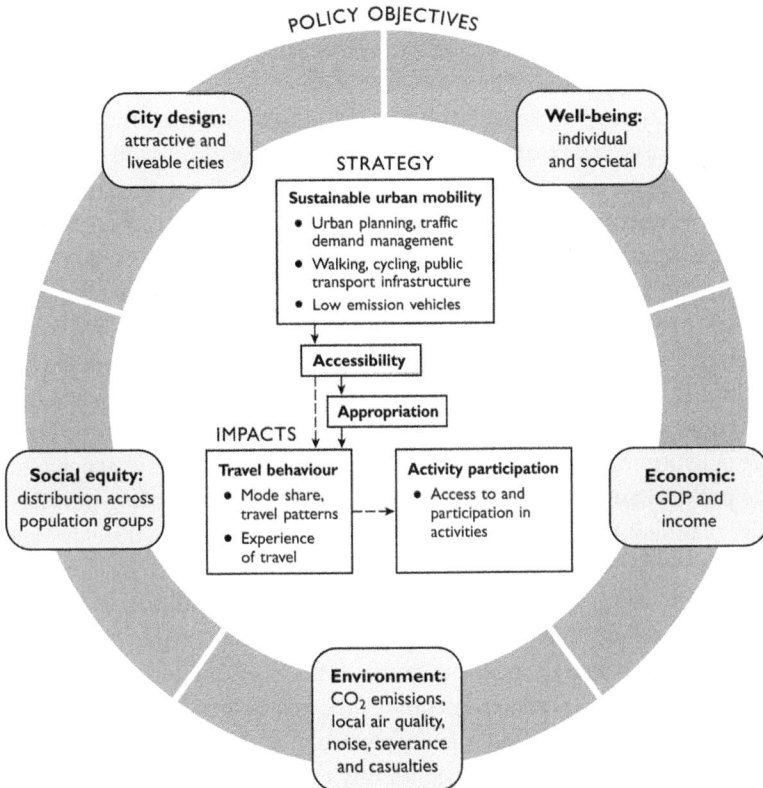

**Figure 2.2** A framework for sustainable urban mobility.

public policy goals. There are problems with contestation and inconsistency in planning over projects, particularly in the implementation stages, leading to delay and abandonment. The issue of subjectivity in relation to transport is poorly understood. We fail to consider appropriation, for example, why some projects (including improved accessibility) are used, and others not.

## Discourse and representation

Interpretative analysis can be used much more centrally in transport planning, with qualitative approaches used to consider the current state of and potential for transport systems, together with motivations for participation in different activities, including through various means of travel. In particular, there is often little problematisation of current transport systems and travel behaviours. There may be little discussion or consensus for the strategic approach and the choice of projects. Assumptions of shared values and aspirations are often misplaced.

Discourse analysis is useful in helping to understand the multiple socially constructed realities found in specific contexts, instead of a more conventionally perceived single 'truth' or reality. This helps us understand the accepted and dominant narratives on specific topics. We can consider how particular viewpoints and positions gain validity over time and space, including how these are organised and produced. Foucault (1966) provides the inspiration for the particular approach of discourse analysis that is utilised, seeking to understand transport as a 'genealogy', including the transport systems, projects and urban developments, and also reflective of the historical trajectory, the political contestation, and the constitution and application of knowledge – all of these specific to the political and cultural context. This broader understanding of transport planning helps us to more fully assess why different strategies are adopted in different contexts, alongside the contestation by different actors, and conclude why particular approaches win and become dominant. Discourse analysis hence provides a framework that allows an understanding of how transport systems and travel relate to individuals and society. This includes the maintenance of transport systems through the status quo, or the repudiation of particular approaches and the emergence of new ones, through discontinuity and rupture. In comparing different transport systems, we can also seek to understand the unseen and unheard, such as what is understood in specific contexts relative to what else might be possible. This includes what is deemed as 'inappropriate', 'undeliverable' or even 'deviant' in specific contexts. The result is that the accepted discourse determines us as subjects and our experiences in the city.

There are different definitions of discourse and indeed approaches to analysing and interpreting discourse empirically; discourse analysis is itself a diverse field. The premise for much of the analysis is to understand how different actors seek to establish a particular narrative or version of events as a means to pursue their own view of transport planning, including their own particular interests. Discourse analysis typically involves examination of conversations, speeches, articles and statements. This assesses narrative structure and phraseology, such as that surrounding the planning or implementation of a new project or strategy. This can be viewed as the discourse of text (Sharp and Richardson, 2001) and is undertaken in a 'bottom-up' manner, using the text as the unit of analysis. Critical interpretations of discourse analysis go beyond the language, aiming to explore the relationships between language and social practice. This involves an interpretative discussion of the narrative relative to wider sociocultural issues. These factors may include political, economic and cultural systems and beliefs, including social movements and changes in cultural norms. Brown and Yule (1983, 1) suggest that

> The analysis of language in use ... cannot be restricted to the description of linguistic forms independent of the purposes or functions which these forms are designed to serve in human affairs.

Keller (2013) advises that this involves consideration of multiple factors, including the actual use of written or spoken language, images or other symbolic forms of social practice. This can include the use of signs and meanings, which themselves are socially constructed and constituted within a wider social context. Fairclough (1992a) has been instrumental in developing this critical dimension to discourse analysis, labelling his approach critical discourse analysis (CDA), to differentiate it from more language-focused content analysis. Fairclough (1992b, 8) states that 'discourse constitutes the social', including 'knowledge, social relations and social identity', and that 'discourse is shaped by relations of power and invested with ideologies'. He views individual interpretation, via the language given, as part of a wider, more comprehensive discursive structure, including the associated historical, sociocultural and political context. There are three dimensions: first, the description of the text; second, the interpretation, in terms of interaction between the elements; and third, the explanation in viewing the text as reflecting and influencing social practice. Textual analysis therefore becomes discourse analysis when considered as part of the social and/or historical context (Wodak and Meyer, 2001, Fairclough and Fairclough, 2012). The relationships are also multi-directional, as discourse does not simply reflect or represent social entities, but also helps shape them. Hence, language and society can be viewed as mutually associated; language can be constituted (be part of and reflect society), but also be constitutive (help to shape society). This critical approach to discourse is reflected by Dryzek (1997, 8), suggesting that discourse represents

> A shared way of apprehending the world, enabling those who subscribe to it to put bits of information together in coherent accounts. The assumptions, judgements and contentions on which each discourse rests provide the basic terms for analyses and debates.

Similarly, Hajer and Versteeg (2005, 175), interpret discourse as

> An ensemble of ideas, concepts and categories, through which meaning is given to social and physical phenomena, and which is produced and reproduced through an identifiable set of practices.

The relationships between language and social practice are produced and organised by situational, institutional and social structures, and change over time. There are three parts to the constitutive relationship: discourse contributes to the making of social identity and the subject position of the individual, there is a construction of social relationships between people, and of systems of knowledge and belief (Fairclough, 1992a). The focus on the critical element involves a normative element, in that there is commentary on what is deemed 'right' or 'wrong' in society and how this might be resolved. The critique assesses what exists, what might exist and how this might be achieved (Fairclough, 2010).

The rationale for using discourse analysis in transport planning is that conventional research approaches focusing on understanding existing travel behaviours or the impacts of new strategies or projects are sometimes simplistic in that they fail to understand the messy processes, complexity and contestation of policy planning and implementation. Dimensions of power, knowledge and ideological conflict are overlooked, but can be very important in the production of transport systems, that is, why and how transport systems are planned and experienced. These issues can be further understood by examining the discourses of strategy and project development, including the different viewpoints of key actors, leading to a critical commentary on the planning and implementation of strategies and projects. Language is central to this, illustrating why certain decisions were made, how they were presented and arguments were won. As Fairclough et al. (2004, 2) explain 'People not only act and organise in particular ways, they also represent their ways of acting and organising, and produce imaginary projections of new or alternative ways, in particular discourses'. Examining discourses associated with particular strategies and interventions allows an understanding of the obduracy of existing structures, practices and the wider habitus, and the tensions between continuity and discontinuity. There is a critical position taken on 'truth' and knowledge and existing positions can be challenged. Where practices are considered to be unjust, they can be reshaped and new trajectories pursued. Further, concepts such as 'sustainable transport' are often interpreted differently and strategies and projects are contested;

hence, discourse analysis allows a greater understanding of the different trajectories and positions taken.

Empirically, there are many and varied approaches to analysing discourse, including conversational and linguistic analysis (such as lexical, grammatical and semantic analysis), conversation and interaction analysis, thematic analysis, narrative analysis, studies of institutional patterns of knowledge and power, critical discourse analysis, political discourse analysis, sociology of knowledge (SKAD) and more mixed qualitative-quantitative approaches such as Q methodology (Keller, 2013). The analysis can be undertaken at the local scale, such as examining the conversations between people, or at the strategic or abstract scales, such as examining the status of a particular social practice. The latter may be more observational in approach, using interviews and related research, rather than being based on the available written concourse. It is this type of observational discourse analysis that is used in this book. There are examples of discourse analysis within the domain of urban planning. For example, Healey (1997) considers the discourse of urban policy, describing a policy discourse as 'a system of meaning embodied in a strategy for action', whilst Mazza and Rydin (1997) examine how urban policy is conceptualised differently across different contexts. Transport planning makes much less use of discourse analysis, but there is some emerging work, which is gradually developing into a distinct research field (such as Flyvbjerg, 1998, Guiver, 2007, Rajé, 2007, Dudley, 2013, Hickman et al., 2021a, Hickman and Hannigan, 2023).

Foucault saw himself as a scholar of the history of systems of thought, hence focused on understanding the emergence of social practices. He saw discourse as a certain 'way of speaking' within the wider structure of society (Foucault, 1969, 193), including what is 'already said' and 'not said' (Foucault, 1969, 26). His work on mental health, imprisonment, sexuality and other topics showed how accepted viewpoints can be challenged and revised over time. For example, in examining the history of the treatment of insanity, Foucault (1961) explains how insanity was deemed 'eccentric' in the Middle Ages, becoming 'embarrassing' in the sixteenth century, and later 'intolerable' in the eighteenth century. His analysis led to a reinterpretation of the role of institutions and changes in the treatment of mental health. There is much scope for such critical interpretation concerning the status of social practices in transport planning and travel behaviours, relative to what is produced and experienced, what might be discussed and not discussed, and how the practices might be improved. Hence, in this way, discourse is interpreted as 'a number of statements for which a group of conditions of existence can be defined ... a fragment of history, a unity and discontinuity ... posing the problem of its own limits, its divisions, its transformations' (Foucault, 1969, 131) and as 'practices that systematically form the objects of which they speak' (Foucault, 1969, 54).

Foucault examines social practices as they evolve over time, including the roles of individuals, institutions, politics, ideologies and particular ways of thinking. He examines texts as part of this, but the focus is on social practices, hence discourse analysis is not necessarily only text-based and is undertaken in a 'top-down' manner, examining the form of dominant practices. He seeks to demonstrate the need for social change, understanding practices in different domains (Foucault, 1961, 1975, 1978a). These are viewed as socially constructed realities, beyond the conventional identification of a single 'reality' or 'truth'. This helps us to examine and expose social structures; indeed, the motivations are often to highlight issues of power or social inequity. Dominant practices are understood as the exercise of power by a particular group, resulting in adverse states such as social inequity (van Dijk, 1993). Dominance may be enacted and legitimised through text and the spoken word. Further, Wodak (2001) highlights that CDA is concerned with analysing opaque as well as transparent structural relationships of control, including where these are manifest in language. Some discourses dominate others, that is, they become the 'mainstream' viewpoint or 'worldview' in a particular order of society. Other discourses can be marginal, oppositional or viewed as 'alternative' discourses (Fairclough, 2010). Hence there is a privileging of some forms of knowledge over others; and knowledge itself is produced and exchanged as a socially constructed reality or discourse (Hajer and Versteeg, 2005). The dominating discourse reflects the habitus, that is, the social and economic conditions that shape the social practice.

The case studies in the book are analysed in Foucauldian terms, examining the planning and development of transport systems and associated travel behaviours in different contexts. Foucault gave no clear methodological framework that could be easily applied in different domains. But, his approach starts with problematisation, that is, defining a problem within specific social practices, and then critically examining the legitimacy of established assumptions, structures and social dynamics in relation to this. Hence, the understanding of history is used to diagnose the present (Kendall and Wickham, 1999). There are some 'key concepts' utilised by Foucault (1969, 1975, 1977, Shumway, 1989, O'Farrell, 2005), or a 'box of tools', that have been drawn together to provide the conceptual framework for this book

**Figure 2.3** Key Foucauldian concepts in discourse.

(Figure 2.3). This assists in reconsidering the practice of a discipline, in this case, transport planning as a socially constructed practice. We can seek to understand why some transport strategies and projects are produced in some contexts and appear impossible in others, including dimensions such as power and knowledge. Transport systems and travel behaviours hence can be considered as part of wider political and cultural practices, delimiting and conditioning norms, thoughts and actions.

The key features of discourse (Foucault, 1969, Shumway, 1989, O'Farrell, 2005) include

- Discursive formation: the group of statements, concepts and knowledge at a given time, such as found in a discipline. The statement forms part of the formation, as a phrase, proposition, act of speech (the part as distinguished from the whole).

- Discursive meaning: the meaning of the statement depends upon the context within which it emerges, such as beyond the language.
- Discursive practice: the historically and culturally specific set of rules that define activities and knowledge within a given period or social condition, including techniques, effects and scientific practices.

Hence, discourse includes forms of representation, presentation and interpretation, in relation to wider political, socio-economic and institutional factors. Discursive practice contributes to reproducing society, including social identities, relationships, knowledge and belief.

## Dimensions of discourse

Discourse is reflective of the social order of the time, drawing on the historical context. There are key dimensions of discourse (Foucault, 1969, Shumway, 1989, O'Farrell, 2005), and these may overlap, including

- History: the order underlying any given culture at a given period of time. Interpretations of history are used to diagnose the present; indeed, solutions provided by past societies may not be the answer to current problems. Archaeology is a term used by Foucault, from the 1960s onwards, to distinguish his work from mainstream historical analysis. He is not so interested in chronological and perceived 'objective' history, but instead focuses on multiple artefacts from a point in time, seeking to understand how these fit together and produce discourses, including accepted viewpoints and knowledge. Hence, he develops an approach to critical archaeological analysis, incorporating issues of emergence, acceptance and transformation (Foucault, 1969). The discursive event, for example, can constitute the present and ourselves, including knowledge, practices, types of rationality, relationships to ourselves and others (Foucault, 1983a). The episteme is viewed as the unconscious structure underlying the production of scientific knowledge in a particular time and place, for example, the paradigm. The event is important as a component of history, but also as a designation of history, in terms of what is remembered and discussed.
- Truth: an event that takes place in history, rather than something that exists, waiting to be discovered. Truth is often used in relation to the subject, power and knowledge, with truth being a category of power (Shumway, 1989). Techne is used to represent knowledge and

know-how. Normalisation is the process of making a particular truth become evident, which, for example, could cover the normalisation of discipline and power. The regimes of truth are the mechanisms involved in the production of discourses, which function as true in a particular context, for example, the science, education, advertising and so on that reinforce the accepted discourse (Foucault, 1975).
- Power: a network of relations between actors; hence viewed as a relation rather than a substance. Power can be productive and repressive, operating through the state and also the wider social body. There are many different manifestations, including the sovereign state (the power of the national, city government or monarchy) and disciplinary (through procedures, rules and regulations). The use of knowledge is important to power, with knowledge viewed as an event and historically derived in nature, rather than innate. Power knowledge is used to illustrate how powerful actors define and apply knowledge to suit their own ends. Forms of knowledge may include working practices, scientific knowledge and the ordering and organising of data. A distinction is made between *savoir*, as the general knowledge; and *connaissance*, as the interpretation of the knowledge, including as a corpus or a discipline (Foucault, 1969). The apparatus (*dispositif*) is seen as the institutional, physical and administrative mechanisms or knowledge structures, which maintain the exercise of power within the social structure; whilst governmentality is the way of administering the population, including the individual and wider cohorts. Genealogy is the term Foucault uses to describe his historical analysis in the 1970s, progressing from archaeological approaches, including dimensions of power and knowledge, the role of key actors, institutions, people and the creation and legitimation of discourses. Biopower is the management of the population, including through disciplinary power (Foucault, 1978a).
- Discontinuity: the break and difference over time. The initial stage is to examine practice, for example, the conventional or expected way of doing something; then, if required, to problematise the practice by developing an understanding of the issue that requires a different solution. Foucault, hence, views problematisation as a consideration of how a field of experience or social practice, once deemed as unproblematic and accepted, comes to be viewed as problematic, including through discussion and debate. This may lead to a crisis in the previous silent behaviour and a revised set of practices (Foucault, 1969).
- Ethics: the relation one has to others or oneself, for example, the moral subject of his or her actions. Experience is viewed as the interrelation

between knowledge, normativity and subjectivity in a particular context and time. Exclusion is the situation and process by which societies exclude certain groups and individuals (Foucault, 1975).
- Subjectivity: the quality of being based on or influenced by individual feelings or opinions. The subject is an entity that is self-aware and capable of choosing how to act. Culture is seen as the organisation of values, accessible to everyone, but at the same time acting as a mechanism for selection and exclusion. The purpose of government is viewed as the conduct of conduct, or the power to influence the actions and behaviours of others (Foucault, 1991).

A clearer understanding is required of the nature of social interactions within society and how these influence transport systems and travel behaviours. Social conflict is not usually settled by the interventions of planners or transport planners acting within principles of rationality and political neutrality (Simmie, 1974). On the contrary, the discourse of transport is one of uneven power, shaped over decades by powerful actors, accepted by governments, politicians and an often unaware and malleable public, captured by their own ignorance and sometimes corruption. There is an uneven exchange between the modes, where national governments can spend billions on highway 'improvements', yet fail to support even basic public transport, walking and cycling facilities, and public space enhancements. The public are often complicit, supporting what is provided for their transport systems. Some continue to insist that they need a car to facilitate their everyday behaviours. Even governance frameworks are subverted, so that public transport is delivered by private operators, under a market-based and competitive basis, and the system is even expected to make a profit, as if this is related to or will lead to high-quality public transport systems and high patronage. Whereas highway development is heavily subsidised, with increased traffic volumes put forward as critical to national economies and other public policy goals. There are positions taken concerning the domains of the political (for example, in the coalitions and choices being made), intellectual (the dominant sciences and procedures) and cultural (orthodoxies, norms and values), all of which remain unchallenged. Look at your city, neighbourhood and street, and we can see what is winning – it is usually motorisation.

Ideas, cultures and histories can hence only really be understood in terms of their relation to power, often involving exploitation (Said, 1978). This book considers these uneven relationships between different transport modes, and the shape of built environments around transport systems. There are important issues of power, domination and hegemony, with the powerful actors often capturing politicians and significant cohorts of the public and maintaining their interests through the resulting infrastructure. The narrative of unsustainable transport is widespread, manifest through traffic reports on the TV or radio, and through books, film and advertising and wider media. Satisfaction from car driving is real for some, yet when many people drive, this positive and conscious experience erodes. Cities, urban areas and their surrounding regions are motorised, not because this is the most efficient mode of travel or approach to city planning, but because it suits some individuals and many vested interests that can derive profit from the process.

## Contestation and appropriation

Two further concepts are important to consider alongside discourse analysis and these are discussed in turn. First, some emerging transport and mobilities research can be better applied in transport-planning practice. This allows us to consider issues beyond the development of infrastructure, which may also be important to changing travel behaviours. There are three dimensions to urban mobilities (Kaufmann, 2002, Kaufmann et al., 2004, Kaufmann et al., 2018):

- The field of possibilities, such as the transport networks available, conditions of access, activities available and the institutions that govern activity.
- Aptitude for movement (motility), including physical capacity, revenue, knowledge and aspiration.
- Movement, which is the actual movement in space, involving physical and electronic interaction over time, for people, ideas, objects and information.

Within this, motility is a key concept that can be viewed as 'the capacity of entities to be mobile according to their circumstance' (Kaufmann et al., 2004, 750). Hence, new infrastructure may result in mobility, or not, depending on the actual takeup of the new project and accessibility that has been provided. This can cover spatial mobility (the movement of people, goods or information from origin A to destination B) or social mobility (the transformation of the social position and distribution of resources for individuals, families or groups within society). Motility is a result of dimensions of access, competence and appropriation (Figure 2.4).

Access issues may include the range of transport networks and communications available, and the spatial distribution of the population, activities and infrastructure. Competence issues include physical ability (such as the ability to move), acquired skill (such as licenses and permits, knowledge of terrain) and organisational skill (such as information). Appropriation issues include aspirations, needs, values and habits. Therefore, the actual use, or not, of mobility, and the reasons for this, become central to understanding the performance of transport systems.

Levels and types of motility will vary in different contexts, reflecting levels of access to different modes, but also how individuals act on perceived or real accessibility. In this way, motility can be viewed as a form of capital, giving access to other forms of capital, such as economic, human, cultural and social (Kaufmann et al., 2004, Shliselberg and Givoni, 2018, Shore, 2023). There are also emerging and important social equity dimensions, as travel behaviours change, reflecting sociodemographic changes (for example, globalisation, ageing societies), emerging modes (shared access to vehicles, bicycles, e-scooters, the public transport and cycling renaissance) and changing trip types (such as longer distance travel and changing commuting patterns). There will be different levels of access to these trends, perhaps with increasing inequity between income cohorts and wider groupings. These issues of appropriation are very poorly understood in transport, but, of course, are critical in attempting to understand the effectiveness of transport systems. The concept of motility can be used to measure progress against sustainable transport aspirations, reflecting that it is not only access dimensions that need improvement – the gains from transport are wider than increased mobility and even activity participation; they also lead to increased human capital of different forms.

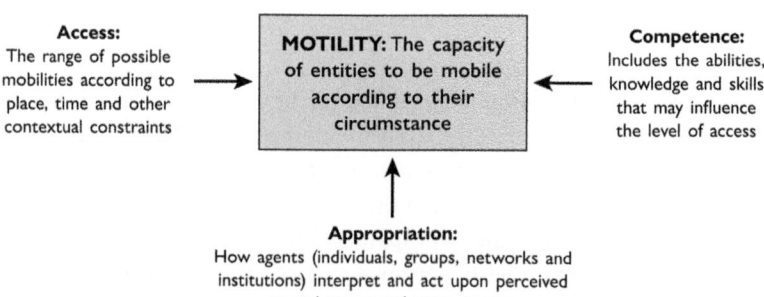

**Figure 2.4** The motility concept. Source: based on Kaufmann et al., 2004.

Second, as Lefebvre (1974) suggests, space is produced, rather than simply existing; that is, it is a product of contemporary social structures and the power relationships within these. This includes the transport systems and streets that we use. The concept of the 'spatial triad' (Lefebvre, 1974, Leary-Owhin, 2016), if applied in transport planning, allows an understanding of what transport systems are presented as the 'conceived space', typically as infrastructure from the city authorities and project promoters. Alongside, experiential elements can be examined, including 'perceived space', which concerns how transport systems are understood by users. Finally, 'lived space' concerns how transport systems are experienced (Figure 2.5). This helps us to examine contestation over the development and use of transport systems. Lived experiences, in this case concerning travel behaviours and activity participation, may be different to what was intended by the project promoters. These experiential elements can be given a much greater focus in research and practice.

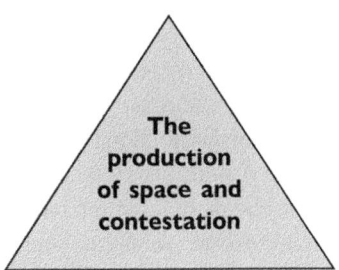

**Conceived space (representations of space):**
The 'official narrative' of plans, images and technical knowledge from city authorities, city planners, transport planners, developers and consultants –
'what is presented'

The production of space and contestation

**Perceived space (spatial practice):**
The built form and infrastructure, everyday routines; giving the official narrative meaning –
'what is perceived'

**Lived space (spaces of representation):**
Spaces as lived and experienced, sometimes counter to the official narrative, leading to counter space –
'what is experienced'

**Figure 2.5** The spatial triad in transport planning. Source: based on Lefebvre, 1974, Leary-Owhin, 2016.

## Case studies and discourse

Narratives are developed for the case studies to help understand the transport and urban development practices found in different cities. The cities are purposively chosen to cover different types of transport systems and contexts internationally. Critical commentaries are used to highlight the positive and negative impacts of the transport systems and also the difficulties in application – to look beyond the promotional and uncritical discussions that are often evident. The case studies are, therefore, more than discussions of good practice. They are discussed in relation to their specific discursive formations and the Foucauldian dimensions of history, discontinuity, power, truth, ethics and subjectivity, and wider related discursive concepts. Each city is described in relation to a specific discursive concept that it most closely reflects, though of course there are wider features that are evident. In this way, the discursive elements are discussed as a comparison across the case studies, rather than for each specific transport system. The case study approach helps provide a detailed and rich understanding of how transport systems and projects have been produced, giving insights across complex phenomena (Flyvbjerg, 2011).

The selected case studies (Table 2.1, Figure 2.6) reflect a specific choice, and there are feasibly many other examples that could give important lessons. Even the mapping of case studies is a representation, telling a particular story. For example, the frequently used Mercator projection presents countries near the Equator as too small relative to their actual spatial sizes and, particularly, when compared to those of Europe and North America. The reason for this is that the map was produced in the 1800s for nautical purposes and presented orbital routes as straight lines. The map became well used and has produced a specific representation of the globe. Other map projections can be used which correct this and show Africa, South America and Asia as nearer to their actual area sizes. Hence, even the choice of mapping base reflects the centrism of research in European and North American contexts – and can shape people's views and interpretations of the world. Figure 2.6 uses the Gall stereographic projection, which gives a mixed projection by area and conformality.

Each city has its own particularities and context, reflecting the problems and opportunities faced. There are potential possibilities for policy transfer, but the focus of the book is more to understand the specific discourses on transport in the selected cities and how they have come to be. Contextual factors affect how transport systems are produced (Robinson, 2015), hence policy transfer is not always as straightforward as might

Table 2.1  Discourse and international case studies

| Concept | City |
|---|---|
| **History**: the order of motorisation given to the rebuilding of the city from the 1950s onwards. | Plymouth |
| **Truth**: an event that takes place in time, that becomes a 'truth' for transport planning in wider contexts. | Oxford |
| | Freiburg |
| | Singapore |
| | Bogotá |
| | Houten |
| **Power**: a network of relations between actors, both productive and regressive. | Chongqing |
| | London, King's Cross |
| | Rio de Janeiro |
| **Discontinuity**: the break and difference over time, different to the expected way of doing something. | Utrecht |
| | Copenhagen |
| | Malmö |
| | Dar es Salaam |
| | Delhi |
| | Shenzhen |
| **Ethics**: the relation to oneself or others, including through inclusion or exclusion. | Manchester |
| | Valenciennes |
| | Medellín |
| **Subjectivity**: being based on or influenced by individual feelings or opinions. | Portland |
| | London, Ealing LTN21 |

be envisaged. As Peck and Theodore (2010) suggest, the movement and application of ideas from one context to another is not straightforward, and the result is often policy mutation, with unintended impacts, rather than simple transfer.

Further, there are critical elements of generalisation in transport planning. First, transport-planning practice has frequently been understood and prescribed in a templated manner, with limited consideration of the context. The classic example is transference of highway-planning practice from North America to wider contexts, irrespective of applicability for different cities. Second, practice in North America, Europe, South America, Asia or Africa, or the so-called 'Global North', 'Global South', 'Developed' or 'Developing World', is often understood in generic terms. Though there are some similarities in transport systems and built environments, these categorisations are simplistic and have little meaning.

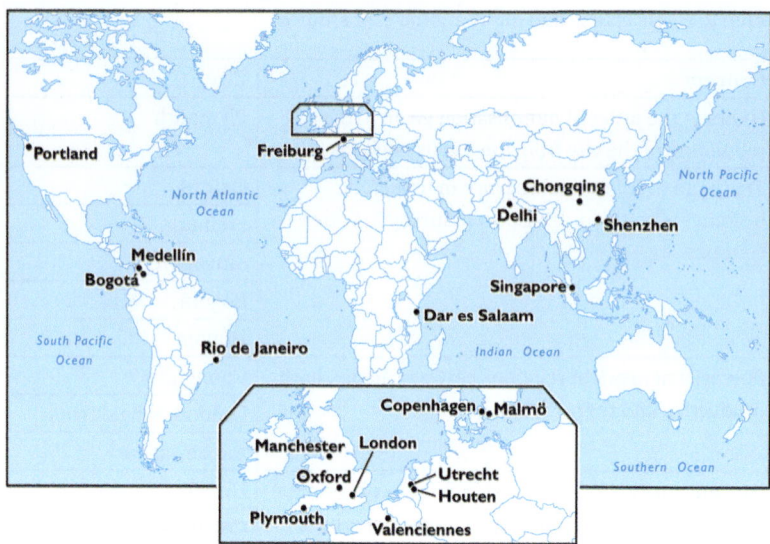

**Figure 2.6** Case studies.

Countries differ hugely within a region and cannot effectively be grouped together under one category. Cities within a country, and even neighbourhoods within a city, can vary dramatically, including by transport systems, socio-demographics, cultural norms and travel experiences. Travel behaviours will vary greatly at the individual level even within cities, let alone at regional scale. Hence, we should adhere to the lessons from development studies, such as from Said (1978) and onwards: we should be aware of the problems of generalisation, including through Western centrism, of the depiction of livelihoods and behaviours with negative stereotypes and of the transferal of 'modern' approaches to contexts where these do not fit. Much of this is simply self-affirmation and even exploitation on behalf of so-called developed countries. This is very clearly seen in the implementation of the North American-inspired motorisation approach implemented across different cities. This generic approach has not worked and has led to very significant adverse impacts. Instead, we should develop transport strategies based on careful analysis of specific contexts and knowledge should be derived from the local context itself. This may often involve participation and co-production of transport systems. This is what we can learn from the different case studies – that there are very different potential trajectories to follow, but these are dependent on the particular discursive structures, of what is possible, in the specific contexts.

## Method of research

A range of approaches were used to generate the material for each of the case studies. Over the period 2014–24, 20 cities were visited, each usually involving between one and three transport projects and/or urban developments. Interviews ($n=150$) were carried out with key actors involved in the case studies, often with discussions during field visits and/or with follow-up discussion. In addition, project reports and academic papers were reviewed for each case study, where available, to provide additional context for the case studies. Two of the case studies drew on additional semi-structured interviews with key actors and the public, carried out in parallel research ($n=18$ in Houten; $n=15$ in London LTN21) (Hickman and Afonin, 2022, Hickman et al., 2025). This helps to further understand the travel experiences in these locations. The interviews help provide a deep understanding of the projects and the underlying issues (Silverman, 2013), revealing processes of planning and implementation. The analysis was primarily inductive, grounding the research in examination of the projects and the inferences drawn from the discussions. However, there is use of a deductive approach in examining the case studies through the Foucauldian discursive framework. Each case study was chosen to illustrate a particular innovation in practice and to represent an element of the discourse framework.

Much of the research was funded by four research projects. Mostly, the case studies were developed as part of the online course on Transforming Urban Mobility, with Deutsche Gesellschaft für Internationale Zusammenarbeit (GIZ) (2019–21); and the online course on Transport and Sustainability for the UK, with the UK Department for Transport (2021–2). Additional case studies were developed during the Sintropher project in Northwest Europe examining public transport provision in peripheral areas (EU, Interreg IVB, 2014–7); a recent study visit with Hunan University, Foreign Expert Programme (2022–3); and wider study tours.

## Structure of the book

Discourse analysis is most often used to analyse a dominant narrative relative to a particular topic at a specific point of time. Foucault, for example, examines the complex exercise of power and knowledge, and brings these into view concerning problematic practices that should be discontinued. This book follows this approach, but also uses discourse from a different perspective. Motorisation and the associated dispersed built environment are considered as the dominant narrative in most cities, but alternative pathways for transport systems are examined, such as varied forms of public transport, walking, cycling and the built environment, to highlight the diverse trajectories and narratives that are possible and the benefits that can be achieved. The private car is often presented as the embodiment of mobility, but I challenge this to show there are very different trajectories available. Hence, the case studies are used to demonstrate the emergence of varied discourses and the potential for discontinuity away from motorisation: that different transport and city-planning pathways are possible relative to the specific cultural context. Within each transport system, there are discourses of dialogue, contestation and dominance, with some narratives gaining prominence over time. In the end, a greater knowledge of the possibilities on offer will help in considering which transport systems to plan and implement in specific contexts.

# Part II
# History

## Challenging the assumed social order

Historical analysis is used to examine and analyse past events, usually seeking the most appropriate narrative to explain patterns of cause and effect. Foucault examines different social practices (Foucault, 1961, 1975, 1978a) to illustrate and challenge the particular social orders at a given time. He examines the limits of systems of thought, institutional practices and contests claims of universal truth (O'Farrell, 2005). Archaeology is used as a concept to introduce a critical view on historical analysis (Foucault, 1969), that is, to explain how a cultural phenomenon came to be, but also to problematise this and challenge the accepted positions and argue for changed social practices. Hence, an examination of history is used to diagnose the social practice under consideration. Latterly, genealogy was used to include dimensions of power in the archaeological analysis (Foucault, 1978a), helping to further consider the values present in contemporary society. The analysis is carried out to illustrate the possibilities for discontinuity and suggest new trajectories, that is, to break the current hegemonic positions and the processes of inertia. Episteme is seen as the framework of unconscious structures underlying knowledge and thought. The event covers more than might conventionally be understood by the term, that is, in this case, an occurrence happening in time, but including human activity, experience and even cultural forms, as well as differing interpretations of these. The period of motorisation, for example, has developed and ordered its own knowledge, to represent and further itself, and these perceived universal claims can be reexamined and repositioned. These issues are examined by considering the regimes of motorisation and the impacts of motorisation; and the history of Plymouth, as an example of a motorised city in the UK.

# 3
# The regimes and impacts of motorisation

## Shaping motorisation and space

The emergence and rise of motorisation and the wider system of automobility are fascinating and central features of contemporary society and globalisation (Sheller and Urry, 2000). Motorisation is seen as a critical organising factor within modernity, including a set of political institutions and associated organisations that regulate and shape transport systems. The transport system hence involves the shaping of rationality, including facilitating the means of travel and the built environment, but also the procedures of urban and transport planning, and the interpretations of how transport planning should be produced. The trajectory of the emergence of motorisation varies in different contexts, with differing narratives, but has had remarkable success in influencing travel behaviours.

Globally, the level of motorisation is running at unprecedented levels across almost all contexts. There were 1.6 billion motor vehicles globally in 2020, including 1.2 billion passenger cars and 410 million commercial vehicles (Hickman et al., 2017b, OICA, 2020). This is an increase in motor vehicles of 4 per cent from 2015–20 and follows continued increases over decades. There are many differences in motorisation rates for different countries in 2020 (Figure 3.1), with the global average at 209 vehicles/1,000 population. Some Western industrial countries are experiencing slowing growth rates or marginally declining car ownership levels, but aggregate motorisation levels are increasing globally and particularly in commercial vehicles. The USA still has the highest level of motorisation (860) alongside New Zealand (869), followed by Australia (737), Canada (707) and Japan (612). European countries have high-middle range levels, including Italy (756), Germany (627), the UK (632) and the Netherlands (588). Asia, South America and Africa have much lower levels of motorisation, but these are rapidly increasing, including China (223), Central and South America (176), Africa (49) and India (33).

The discursive formations and practices behind these motorisation rates include the development of highway networks and also a specific set of processes, regulations and rules that define mobility. This includes the emergence of disciplines such as traffic engineering, transport planning, real estate and urban planning, which produce space in a way that encourages transport systems and travel behaviours based around the car. There are processes of delimiting, designating, lobbying and supporting motorisation, giving preferential status to the car as the preferred means of travel, thereby making motorisation implementable as a form of transport. The discursive formation includes the creation of knowledge surrounding motorisation, including associating the use of the private motor car with ideals such as convenience, freedom, privacy, autonomy, movement, progress and status. Through this, the artefacts of motorisation are legitimised, including cars, highways and the dispersed urban form. Böhm et al. (2006) discuss the regimes of automobility, including the systematic aspects and elements of power. The regimes of truth help to produce and reproduce the taken-for-granted character of car driving as the dominant form of travel – it is perceived as what most people do or would like to do. This involves shaping the discursive meaning so that the car is seen as the most effective, efficient and aspirational mode of travel, and the primary mode that cities and regions can be designed around. Increasing levels of motorisation are presented as inevitable, indeed something that we should aspire to. The regime of power is seen in the governance of automobility by motor manufacturers, motor organisations, transport authorities, universities and research departments. Indeed, there are often specific highways agencies that lobby for significant funding and implement new highway infrastructure. The public are also involved, with a significant proportion vociferously supporting individual car usage and highway development, including by contesting attempts to reallocate streetspace away from the car. There is also the power of the vehicle, such as sports utility vehicles (SUVs) and light duty vehicles (LDVs), over other street users, such as pedestrians and cyclists. The regime of subjectivity is seen in the intertwining of automobility with notions of individualism. 'True' mobility is shaped as individual motorisation, with obvious neoliberal influences in terms of the self-interested individual and focus on consumerism. Car driving is understood, by many, as what 'normal' people do in order to travel.

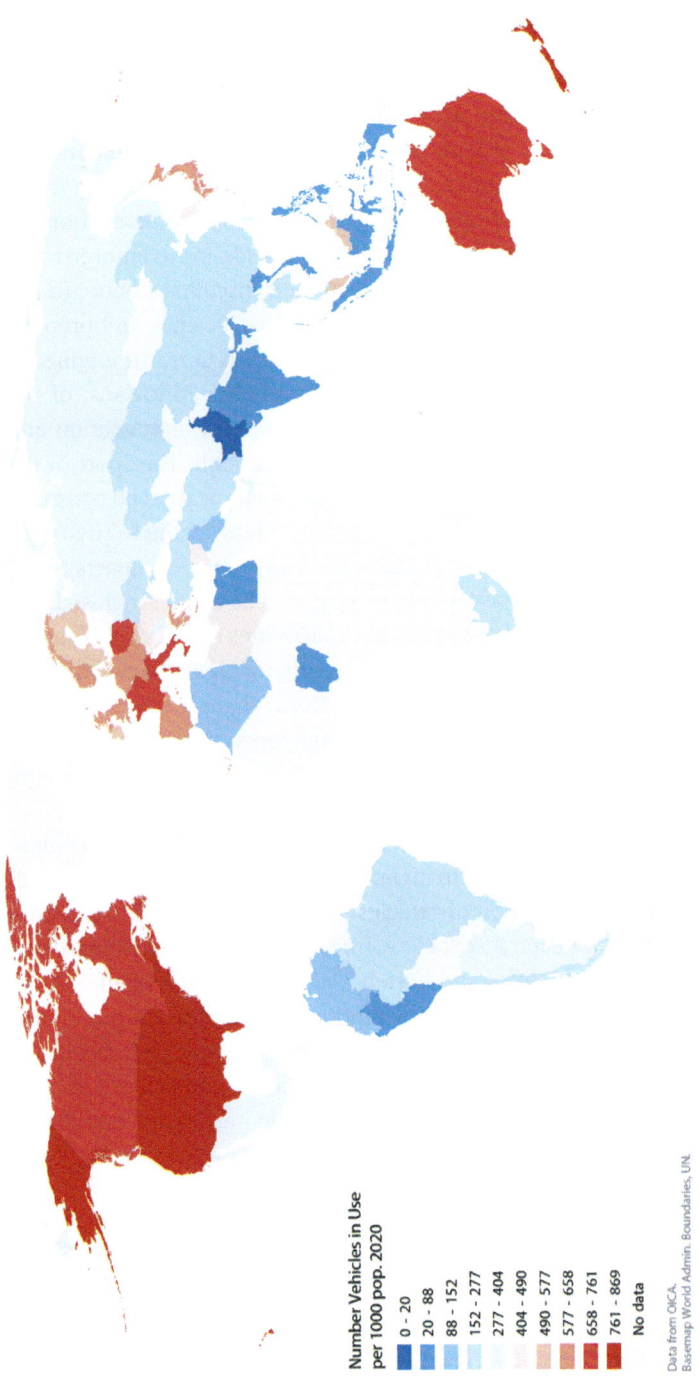

**Figure 3.1** Motorisation rates by country. Source: Hickman et al., 2017b, International Organization of Motor Vehicle Manufacturers (OICA), 2020. Drawn by Duncan Smith, CASA, UCL.

Further, it is not enough to travel in a small, less energy-intensive and low-emission car; instead, advertising pushes people towards large and expensive vehicles. Conversely, pedestrians and cyclists are framed as 'deviant' and bus users as 'failures in life'.

The episteme is seen as the underlying structure and conditions that regulate the forms of knowledge that emerge. The combination of power and truth is important to help shape knowledge, including social practices, rules and procedures, which help to separate perceptions of 'true' and 'false' (Foucault, 1977). The apparatus helps to enforce these regimes of motorisation, including sanctioned discourses and cultural norms, through institutions, laws, administration, science and propositions. This so-called knowledge is deeply ingrained in traffic engineers, transport planners, urban planners and the public. Politicians, of the right and sometimes of the left, consistently support motorisation and invest in highway programmes at the expense of public transport, walking and cycling. Yet the discursive formation is simply a social construction – and could be constructed in a very different manner. The tragic results in many contexts are that there are many adverse impacts at the societal level and to not have access to a car is to be disenfranchised; the car is the only possible way to access many activities (Freund and Martin, 1993). The private car has, in the vast majority of contexts, prevailed as the primary means of everyday transport. Even where there are forms of public transport, or walking and cycling networks, the car is often present, and most likely the investments in alternative forms of transport are not sufficient to develop high levels of mode share.

Hence, there is a strategic notion to knowledge in transport planning: it is developed and used to maintain the power of motorisation and, ultimately, the consumption of vehicles. Private cars, SUVs and LDVs are critical products in capitalist society, furthered through neoliberal governance models, and cities and lives have been designed around them. There are huge conflicts between the products motor manufacturers wish to sell and their adverse impacts on society, and the root problem, as Marx (1887) would tell us, is capitalism. Attempting to achieve sustainable mobility without tackling the context of capitalism and neoliberalism is unlikely to be successful to any significant degree.

## Car advertising

The dominant culture of motorisation is produced and reproduced in many forms, including via films, books and wider media (Sheller,

2004, Smoak, 2007). Alongside, the private car has been heavily advertised over the last 70 years to encourage consumption, and the changing nature of car adverts over time can be seen in print advertising (Hickman et al., 2021b, Hickman, 2023). There has been an increasing sophistication in the messages over time, initially in selling the basic utility of the private car to the mass population, and gradually developing through association with status and success, access to the open road, 'useful' gadgets and in-vehicle safety, to more subtle messages of the car as an assumed 'environmentally friendly' purchase (Figure 3.2). The private car is positioned throughout as the latest in fashion and styling. In the USA, vehicle advertising accounted for nearly $15 billion of expenditure in 2012. Nearly half of this amount was contributed by three companies – General Motors ($3 billion), Ford ($2.3 billion) and Toyota ($2 billion) (Statista, 2016). Hence, this is a major financial undertaking and helps explain the power of the messaging and the development of car culture. The latest advertising aims to position electric and hybrid vehicles as the solution to environmental problems such as climate change. But there is 'greenwashing' of the environmental benefits of low emission vehicles, as the lifecycle emissions including power supply are ignored, are often far from renewable, and the expense of the purchase is still only available for some (Hickman, 2023).

## The impossibilities of motorisation

There are many very significant adverse impacts of motorisation, including energy consumption, carbon dioxide ($CO_2$) emissions, traffic casualties, health impacts of inactive travel and the impact on the built environment (Hickman et al., 2017b). All of these are overlooked in public policy as the motorisation narrative is given precedence. They are sometimes presented as 'indirect' or 'side' impacts, perhaps to help diminish their scale, but they should be viewed as the impossibilities of motorisation – they are so significant as adverse impacts that they make the motorisation dream impossible to implement on a mass scale across multiple contexts.

Increased levels of mobility are leading to huge demand for energy, largely in the form of petrol and diesel to fuel vehicles (Hickman et al., 2017b). In 2015, the world consumed 13,559 million tons of oil equivalent (mtoe), an increase of 35 per cent on 2000 levels (International Energy Agency, 2015). The majority of this energy is now being consumed

**Figure 3.2a** Baker Electrics, 1910s – early association with excellence, elegance, refinement and perfection.

**Figure 3.2b** Ford Mustang, 1960s – sporty and luxurious, and leading to success in life – Mustangers always win.

**Figure 3.2c** Toyota Corolla, 1970s – fully endowed with gadgets, including wheel covers, rear window defogger and reclining bucket seats.

**Figure 3.2d** Land Rover Discovery, 2019 – giving access to the open countryside, and even, improbably, the top of the mountain.

**Figure 3.2e** Toyota Prius, 2010 – positioning EVs and hybrids as the solution to environmental problems. Backgrounded with deep blues, greens and beautiful landscapes to reinforce the environmental themes.

**Figure 3.2f** Audi Q7, 2016 – EVs and hybrids as the ever-so-simple solution to climate change, requiring only a household plug for charging, suggesting that your travel behaviours can remain the same. But this overlooks many issues, such as the need for a renewable power supply, driving and fuelling behaviours, the cost of EVs or low emission vehicles, which is prohibitive to many, and the wider adverse impacts of vehicles, such as traffic casualties and impacts on the built environment, which are still unresolved.

**Figure 3.2** The changing shape of car advertising. Source: Hickman, 2023. Print adverts 3.2A-F courtesy of the Advertising Archives.

by non-OECD countries, that is, non-Western industrialised countries in Asia, South Amercia and Africa, demonstrating the increase and spread of energy consumption globally. Transport energy consumption is a large part of this – in 2015, the transport sector consumed around 2,200 mtoe, equating to 23 per cent of energy usage, growing at nearly 2 per cent per year over the last decade. The vast majority of this comes from oil-based sources (94 per cent) and there has been little change in the dependence on oil for travel since the early 1970s (International Energy Agency, 2016).

Transport is the key sector that is failing to contribute to reduced $CO_2$ emissions, including surface-based travel and air travel (Hickman and Banister, 2014). Even the so-called progressive cities are only reducing transport $CO_2$ emissions marginally. Globally, there is a large difference in national and per capita $CO_2$ emissions. The Emission Database for Global Atmospheric Research (EDGAR) gives data for aggregate emissions, covering the industrial, residential and transport sectors. Since the beginning of the 2000s, global fossil $CO_2$ emissions have grown steadily in comparison to previous decades, mainly due to the increase in fossil $CO_2$ emissions by China, India and other emerging economies (Crippa et al., 2022). The USA, Canada and EU27 have reduced their overall $CO_2$ emissions in recent decades, but only marginally. In 2020, global fossil $CO_2$ emissions decreased by 5 per cent in comparison to 2019, mainly because of the Covid-19 pandemic. However, in 2021, global emissions increased again, almost to the level of 2019, reaching 37.9 Gt, just 0.4 per cent lower than in 2019, with a reversion to pre-pandemic $CO_2$ emission levels. In most countries, even where overall $CO_2$ emissions are decreasing, the level of transport $CO_2$ emissions is rising. Global transport $CO_2$ emissions rose by 65 per cent from 1990–2021, and even in the EU-27 they increased by 16 per cent.

One of the most pervasive and unacceptable consequences of motorisation is the level of traffic-related casualties – a level that is seemingly 'normalised' and overlooked as an acceptable consequence of being able to travel around by private car (Hickman et al., 2017b). The number of casualties is, however, horrendous: around 1.2 million people die globally each year resulting from road traffic crashes and road traffic injuries (World Health Organization, 2023), or around 3,300 deaths per day. Traffic casualties are the leading cause of death among young people aged 5–29. Nearly 50 per cent of deaths are 'vulnerable road users', in other words pedestrians, cyclists and motorcyclists. Between 20 and 50 million more people suffer non-fatal injuries, with many incurring ongoing disabilities. Road traffic death rates (per 100,000 population) vary significantly by country in 2016: the USA has a high rate of road traffic deaths at 12.4 per 100,000 population; Australia (5.6), Germany (4.1), the UK (3.1) and other European countries are relatively low; but there are very high rates in China (18.2), India (22.6), Brazil (19.7) and other Asian and African countries, where traffic safety is poor and death rates can exceed 25 deaths per 100,000 population. In China and India, for example, there were an estimated 250,000 and 300,000 road traffic fatalities in 2018, respectively (Hickman et al., 2017b).

In addition, inactivity, including that related to inactive travel, leads to high levels of non-communicable diseases (NCDs) (World Health Organization, 2015, Hickman et al., 2017b). NCDs killed around 38 million people in 2012, representing 68 per cent of 56 million global deaths. The four main types of NCDs and numbers of deaths per annum are as follows:

- Cardiovascular diseases (mainly heart disease and stroke): 17.5 million deaths.
- Cancers (including endometrial, breast, ovarian, prostate, liver, gallbladder, kidney and colon): 8.2 million deaths.
- Respiratory diseases: 4 million deaths.
- Diabetes: 1.5 million deaths.

Though NCDs are not directly caused by motorisation, poor urban planning, dispersed built environments and a lack of infrastructure for walking and cycling contribute to the problems by making it difficult to choose active means of travel (Woodcock et al., 2009). The most important risks leading to NCDs are high blood pressure, inadequate intake of fruit and vegetables, being overweight or obese (also linked to type 2 diabetes), physical inactivity and tobacco use. Hence, five out of these six risk factors are closely related to diet and physical activity. Obesity is an important risk factor for NCDs and global obesity has more than doubled since 1980, now representing a very significant health problem. In 2014, more than 1.9 billion adults (18 years and older) were overweight; of these, over 600 million were obese. Thirty-nine per cent of adults were overweight in 2014, and 13 per cent were obese (World Health Organization, 2015).

## Urban growth and motorisation

The huge projected urban growth makes increased motorisation across multiple contexts even more untenable. The global population of 8 billion in 2022 is expected to reach 8.5 billion by 2030, 9.7 billion by 2050 and 11.2 billion by 2100 (United Nations Department of Social and Economic Affairs, 2015). This is an unprecedented level of urban growth, with much of the increased urban population expected in the so-called Global South – in Asia, South America and Africa (United Nations Department of Social and Economic Affairs, 2014):

- In 1800, when the global population was around 1 billion, only 3 per cent of the population lived in urban areas, and only one city (Beijing) had a population greater than 1 million.
- In 1990, there were 10 megacities, home to 153 million people, accounting for less than 7 per cent of the global urban population; there were also 21 large cities and 239 medium-sized cities[1].
- In 2014, there were 28 megacities, with 453 million residents, or 12 per cent of the global population; and there were 43 large cities and 417 medium-sized cities.
- In 2030, there are expected to be 41 megacities, 63 large cities and 558 medium-sized cities.

The discursive formation of motorisation ignores all of these very significant adverse impacts. Energy consumption, $CO_2$ emissions, traffic casualties and inactive travel are 'normalised', as if the environmental and health problems are worth the convenience of individuals travelling around in a private car. The discursive practices that follow mean that the street is given over to the car, and many of our cities are ruined as the vehicle becomes too dominant. Yet the regime of automobility is incredibly fragile because of these adverse impacts. Surely, it is only a matter of time before more people realise this is not an acceptable price for individualised mobility, particularly when other, more attractive options are available. It is the scale of urban growth that will make this ever more evident.

## Note

1. A megacity has a population of 10 million or more; a large city 5–10 million; and a medium-sized city, 1–5 million.

# 4
# Plymouth

## Plymouth and Abercrombie

Examining the archaeology and genealogy of transport-planning practice, in many cities and wider regions, illustrates how the private car has been given priority in the built environment over the last 70 years. The strategies for highway planning and resulting rates of motorisation differ, but much space has been given on the street to the car in almost all cities, internationally. Associated with this are very significant adverse impacts; but often these are overlooked as private car usage is perceived as a necessity for life by decision-makers in public policy and also, more widely, by politicians and the public.

There are many cities that could be used to examine the trajectory of motorisation, and perhaps most obvious would be one from the USA. But Plymouth, in the southwest of the UK, is used to illustrate the discursive formation of designing the city around the use of the private car. Plymouth was rebuilt in the years following World War II (WWII), using 'modernist' urban-planning principles to reshape the medieval city centre that had been heavily bombed. Sir Patrick Abercrombie, one of the leading planners of the time, produced the masterplan for the city, the Plan for Plymouth (Abercrombie and Paton Watson, 1943). The city centre bombing destruction was extensive, as Plymouth's naval dockyards were a key target, but this offered an opportunity for comprehensive redevelopment. The key event, in Foucauldian terms, was the reconstruction of the city centre around a new urban highway system, which followed the progressive urban-planning principles of the time. Abercrombie was ambitious and suggested 'a plan which would embody drastic proposals' (Chalkley and Goodridge, 1991, 66). The conceived space, via the masterplan, produced a completely redesigned city centre, developed in Beaux-Arts 'City Beautiful' style. This reflected modernist urban-planning approaches as found in cities such as Washington DC, Paris and New Delhi (Boughton, 2013). The municipal-led planning aimed for a bright, clean and spacious central area and a geometric street network, with a perceived more efficient use of space. This was in contrast to the previously irregular and constrained medieval streets. The Plan for Plymouth was prepared and reconstruction started within 3 years. The city centre was assembled as a single, cleared site of nearly two hundred acres, all of which was brought into public ownership. Separate precincts were developed for retail, offices, culture and civic government, and the central street network was refocused on a grand axis, Armada Way. This replaced the previous medieval street network, providing an impressive vista stretching for over 1 kilometre in length, running from the North Cross roundabout (south of the railway station) to the open space of Plymouth Hoe (the high ground), overlooking the seafront and Plymouth Sound. The Royal Parade acted as the second axis, running east to west, separating the retail and commercial areas (Figure 4.1). The central retail precinct was constructed in distinctive white Portland stone, giving a rare and still intact example of a modernist city centre in the UK (Gould, 2010). The old city centre streets were cleared away and the new centre was surrounded by an inner ring road, of mostly two- or three-lane carriageways, denoting the importance given to direct access to the city centre by car. Beyond the city centre, neighbourhood units were designated as suburban living areas, each with a local centre. These neighbourhoods replaced the poor-quality housing found in the central areas, with the new houses designed around gardens, aimed at improving public health. The premise was for direct and convenient access to the city centre by car and car parking provision was massively increased on the edge of the city centre. The episteme of postwar urban planning reflected the aspiration to improve postwar lives and a confidence in grand-scale, top-down masterplanning. This developed into the newly emerging profession of town planning (Chalkley and Goodridge, 1991). For Plymouth, and many cities, this enabled a dramatic reconstruction of the built environment, but came to be heavily criticised (such as by Jacobs, 1961) due to the displacement of communities, often with social equity and racial dimensions, and reflecting poor participatory processes with the public. The planning process in Plymouth was completed in almost total secrecy, seeking to avoid land speculation, and there was no public participation until the proposals were presented to the council in 1944. There was a generally favourable reception and the only major concerns were over the cost of reconstruction. The new city and street layout in Plymouth demonstrated the huge ambition, but, viewed from today's perspective,

**Figure 4.1** Plymouth city centre and surrounding developments.

the misplaced knowledge of the city planners of the day. Motorisation became the designation of history, and the image of Plymouth that is seen and discussed today. The space given to highways means that there are difficulties, in contemporary times, in moving away from this use of space once the car has been accommodated in the city.

## Post-World War II urban decline

Over a quarter of Plymouth's employment before WWII had been at the Royal Naval dockyards. But, from the 1950s onwards, the unanticipated decline of employment in the dockyards became the key event in the city, leading to economic struggles over decades, and there was little replacement found for the employment loss. Employment levels and incomes remained low for years and much of the urban fabric deteriorated. Plymouth, with a population of just over 250,000, now ranks in the worst 20 per cent of local authorities in the UK for levels of multiple deprivation. Some local neighbourhoods, such as Devonport, rank in the worst 5 per cent, despite significant efforts to regenerate (Plymouth City Council, 2014).

The growth in traffic was also unexpected over the postwar decades, with large increases in car ownership and use. This meant that the new road capacity was soon full and the city centre became congested and polluted with circulating traffic and parked cars. Pedestrianisation of the central precinct in 1987 helped move some of the traffic away from the central area. But this was a difficult project to implement and was decades behind other pedestrianised city centre projects carried out from the 1970s onwards in mainland Europe (Monheim, 1990). The pedestrianisation project in Plymouth was controversial before implementation, yet strongly supported following delivery (Shepley, 1991). The process of implementation of pedestrianisation schemes is often similar: local shopkeepers, businesses and politicians attempt to block the proposals, with concerns that their businesses will suffer. Over time, town centres become much more vibrant with pedestrians, and the antagonists are won over, often becoming fervent supporters of pedestrianisation. Beyond this small pedestrianised central area, the built environment in the city centre and beyond remains heavily dominated by highway infrastructure, with much space allocated to the car, and there are high levels of traffic.

## Responding to motorisation

The economic decline led to decline in the built environment and the emphasis on providing for highway capacity meant other modes were underprovided. The current mode share for Plymouth shows the continued dominance of the car, with the car accounting for 61 per cent, bus 11 per cent, walk 14 per cent and cycle 3 per cent of journeys to work (Office for National Statistics, 2011), even in a city with low average incomes. Where some of the historic city fabric still exists, such as in the Barbican district near the quayside, the management of traffic has been ineffective and traffic movements dominate the narrow streets. Car access is provided directly to the quayside, where space could be better used for public space improvements and outdoor pedestrian space. Bus priority is poorly implemented across the city and journeys by bus are often delayed at key junctions. There are hardly any segregated cycle lanes across the city, beyond a very short, discontinuous route on the Hoe.

The defining image of the priority given to the car over other street users and indeed the urban fabric is given by Charles Church, which was destroyed by fire following the bombing of Plymouth during WWII. The church was left unrestored as a monument to the casualties of the war, but subsequently surrounded by a traffic roundabout, which gives a very poor setting and ensures that pedestrians cannot easily get to see it. Public policy interventions have been completely inadequate in reducing the priority given to car use on the city streets; indeed, there appears to be no real debate or appetite for change amongst politicians or the public. A significant proportion of residents have become reliant on using the car for most of their journeys. The lived experience seemingly supports the space that has been conceived, yet there are many that are excluded from using cars as they do not have access to the car. There appears little demand for transport provision beyond more space for the car, but perhaps this is more a reflection of the 'fixity' of motorisation (Matthies and Klöckner, 2015) when embedded within lifestyles and poor awareness of the different transport systems that could be provided.

**Figure 4.2** Plymouth motorisation and the city. The HISTORY of planning includes the decision to urban plan in a 'modernist' way and to give cars priority – seen as 'progressive' at the time – which led to a particular shape of the city and problematic travel behaviours.

**Figure 4.3** Plymouth highway planning. The city centre gives too much space to the car and car parking – and the result is that the built environment is unattractive, including for pedestrians.

**Figure 4.4** Charles Church and highway dominance. The DISCURSIVE FORMATION is to give priority to the car. This is also reflected in the scientific discipline of transport and urban planning and the application of knowledge in Plymouth – the preference given to the car has ruined the city.

**Figure 4.5** Pedestrianisation in the city centre. Understanding the development of the city, as an ARCHAEOLOGY, helps to challenge the accepted positions taken over time. The pedestrian area was implemented in the 1990s, and has removed traffic and enhanced walkability in the central zone. But, it is still ineffective in the evenings as pedestrian activity is low due to insufficient central area housing.

**Figure 4.6** 'Sustainable neighbourhood' at Sherford. The key EVENT was to redevelop the city when motorisation was perceived as central to urban and transport planning. The discursive practice continues today, continually influencing travel behaviours, human activity and experience. Even the new, supposed 'sustainable neighbourhoods' such as Sherford are very car dependent.

**Figure 4.7** Poor cycling provision in Sherford. There are attempts to encourage cycling, but there are few safe, segregated cycle routes – hence cycling remains at incredibly low levels.

## The discipline of transport planning

The process of transport planning shapes the projects that can be built, reflecting the apparatus of the institutions, administration and legal framework. Science is used to shape a specific set of knowledge, procedures and datasets, which become the expected standards of practice. This discursive practice is particularly evident in the discipline of transport planning, developing from the 1950s onwards, and shaped to help facilitate increased motorisation in the USA and later in wider contexts such as the UK. Some of the earlier publications on transport planning give an indication of the early thinking. Alker Tripp, for example, was Assistant Commissioner of the London Metropolitan Police and responsible for traffic control in London. He represented the traffic engineering approaches that came to be used across the UK, arguing that traffic safety was a matter for the police, that is, protecting lives at a time when traffic casualties were rapidly rising. A code of traffic regulation and control was developed, including road design, speed limits, traffic lights at junctions, pedestrian crossings and guardrailing, and segregated movements, all aimed at improving vehicular flow (Tripp, 1936, 1938). This police-derived and traffic safety origin was often seen in the initial development of the traffic engineering discipline (Norton, 2011). Tripp argued for increased highway capacity and the segregation of fast-moving vehicles from slower traffic and pedestrians. He particularly prioritised vehicular movements over pedestrian and cycle movements:

> Motor traffic will never, and can never, mix safely with pedestrians and pedal cyclists. It is the policy of continuing to mix them that is hindering the proper and orderly development of motor traction. (Tripp, 1936, 83)

And that the blame for congestion is

> ... the careless pedestrian who steps into the road, the rash pedal cyclist who swerves across the traffic, the reckless motor driver who suddenly cuts out. (Tripp, 1938, 119)

This type of thinking, often derived from perceiving vehicular movements as similar to a liquid and requiring improved flow, was influential in early traffic engineering, and indeed remains today. This was furthered by Abercrombie and others such as Buchanan (Ministry of Transport, 1963), who warned about the 'impending motor age' and the need to accommodate vehicles as car ownership became widespread across the population. Buchanan's suggested remedies, again much inspired by practice in the USA, gave too much emphasis to accommodating increased highway capacity, including through mass reconstruction of cities. Even 50 years later, there was similar guidance to prioritise road capacity for the car, such as the influential and long-lasting governmental guidance for the design of residential streets in the UK (Department of the Environment and Department of Transport, 1992). Over decades, transport planning has been gradually developed and refined as a discipline. Transport is conventionally understood as a derived demand, where travel is perceived as a cost to access a range of activities. Hence, there is a focus on increasing mobility and reducing travel time. Even with accessibility planning, which changes the focus to increasing the choice of destinations and activities, transport is still largely seen as a derived demand. There is poor understanding of the appropriation of travel opportunity, that is, how parts of the transport system are used and why. Indeed, the intrinsic value of the journey, whether for activities undertaken whilst travelling, or for the development of human capital, is overlooked. Transport modelling has played a central role in estimating current and projected travel behaviours. For highway planning, this typically involved the four-stage model, where traffic demand is forecast in stages: trip generation to determine the number of origins and destinations (based on demographic factors), trip distribution, mode choice and route assignment (Banister, 2002). A gravity model was and is still often used to estimate movement flows, based on the attractiveness of a zone and distance between origin and destination. More sophisticated modelling approaches are sometimes used, such as activity-based modelling, where the scheduling of activities is examined rather than trips. But, still, the premise is on forecasting demand and catering for this, an approach known as 'predict and provide'. Alongside, there is project appraisal, using cost–benefit analysis and multi-criteria analysis to prioritise projects, mostly premised on assumed cost effectiveness, and largely representing time savings relative to cost (Hickman and Dean, 2018). Hence, the motorisation system preserves the characteristics of its own purpose, developing approaches through which highway investments can be prioritised and funded. Concepts were developed, such as 'cost effectiveness', 'time delay', 'time savings', 'congestion', 'traffic bottlenecks', all of which become deeply ingrained in transport-planning practice, and even amongst the public, but simply serve further highway investment. Wider means of travel are denounced as unconventional or even pathological (such as walking, 'jaywalking', cycling and use of the bus). The discursive practices, hence, modify the domain that they relate to (Foucault, 1969), including with quantitative analysis given precedence over qualitative analysis in transport planning. This is despite the focus for transport planning being on understanding individual behaviours.

The transport-planning process is put forward as an objective, technical exercise, led by the expert analyst. But, within this, there are many normative positions, involving perspectives and values on the particular working procedures, data to analyse, transport projects to fund and the use of space on streets. The transport network is generally conceived as a system, modelled on fluid dynamics, where the volume of traffic can be more effectively spread around the container within which it is held (Newman and Kenworthy, 2015). It was believed that more highway capacity would lead to the smoothing of traffic flow and better traffic conditions, ignoring that there are wider demands for space from other street users, indeed that increased road capacity will lead to more traffic where there is much latent demand, known as induced traffic. The discipline developed within the neoliberalism frame, where economic growth, mobility growth and individualised travel, utilising an expensive product to be consumed, are deemed positive and unchallenged objectives.

Foucault (1969) usefully distinguishes between different forms of knowledge, which are not so well-defined in the English language: *connaissance* (the corpus of knowledge in a particular discipline) and *savoir* (knowledge in general). These are both useful to consider in transport planning. The connaissance, or discipline, of transport planning was established to help support the building of highways and convert streetspace for the use of vehicles (Norton, 2011). The analytical approach has largely been based upon forecasting future demand and then providing infrastructure to meet this demand. The concept of *homo economicus* is applied, assuming that the individual is an entrepreneur of her/himself, seeking to maximise mobility for economic gain (Foucault, 2008). The individual is deemed as selfish, as his own producer and the source of her/his earnings, including choosing the most efficient journey from A to B in terms of time. This framework and process has been applied internationally for decades, remaining largely unquestioned in practice, as a template to be transferred irrespective of the context, and leads to much highway building. But, of course, it has many very significant flaws that require critical discussion and problematisation. A changed approach is required due to the adverse impacts of motorisation. There has been some critique of the predict and provide approach (onwards from Goodwin et al., 1991). In reality, traffic acts more like a gas than a fluid, and the new space given to the car is filled up almost immediately as people make more and longer trips. The opposite, of course, can occur: where traffic capacity is reduced, traffic then 'disappears' as people take other modes or change their trip patterns (Cairns et al., 2002). The intrinsic value of travel has been raised, largely in terms of the possibility of carrying out activities whilst travelling (Mokhtarian and Salomon, 2001, Jain and Lyons, 2007). These fundamental points have been made in research over decades, but surprisingly have not led to more in-depth research studies or significant changes in the practice of transport planning. There have been calls for consideration of other public policy objectives, such as climate change and social equity, and to focus on a vision-led approach rather than forecasting (Lyons, 2012, Hickman and Banister, 2014). However, in transport planning, the focus remains on forecasting traffic demand and improving traffic flow and efficiency, as measured through time savings in project appraisal. This puzzling inertia in transport planning leads to the majority of funding being prioritised for highway investment and little significant investment in other modes. The process involves minimal public participation, with only very limited discussion with wider actors and the public on pre-selected project options. Progress towards climate change and other environmental objectives remains poor and social equity is largely ignored, including the uneven distribution of the use of transport, activity participation and the displacement of lower-income population groups around new transport projects. This conventional process of transport planning – the old paradigm – is particularly not useful when current trends are environmentally and socially unsustainable and significantly different trajectories are required (Hickman and Banister, 2014). For this, a revised transport-planning approach is required, with a different use of quantitative and qualitative analysis, including within transport planning and project appraisal.

The discursive practice is hence very important, in that a historically and culturally specific set of rules and procedures were developed to form the transport-planning discipline, helping to define knowledge in a given period or social condition. There is an ideological shaping and functioning of the science, and the process has been used to develop motorisation. The process is ineffective in supporting public transport, walking and cycling. The wider savoir (knowledge in general) coalesces around this mainstream technical view, assuming that catering for traffic demand and allocating space on the street mostly to the car is the 'best' way, indeed the 'only' way to shape the transport system. In most urban areas, this can be seen with the projection of traffic demand and then the building of highways to ensure motorised access into the city centre, ignoring the impact on the existing built form and existing communities. The street is mostly given over to the use of the private car; consequently, buses, pedestrians and cyclists are given much too little space. This was an integral part of the modernist urban-planning vision and led to the complete reshaping of cities around the use of the car. Any opposition to mainstream support for motorisation faces many difficulties; vocal cohorts within the public often support the continued use of cars for everyday travel. It is proving difficult to reshape transport planning into a process that can more effectively support investment in public transport, walking and cycling, even to the extent that a significant proportion of transport planners, engineers and economists defend the approaches they are familiar with in application.

## Contemporary urban development

Meanwhile, back in Plymouth, the city authorities are attempting to regenerate the city centre and provide additional housing, including city centre and suburban housing, for a growing population. The unique modernist city centre gives many opportunities, with current projects aiming to revitalise and encourage a more vibrant central area. There are recent urban-planning strategies, including David Mackay's updated Vision for Plymouth (MBM Arquitectes and AZ Urban Studio, 2003), and the city population is expected to increase to around three hundred thousand. This will include new urban living and a refurbished city centre and harbour front. Recent major projects include the refurbishment of the civic centre, an expanded Plymouth University and new student accommodation. New retail areas and community facilities have opened, such as Drakes Circus and the Box; a new boulevard, on Millbay Road, will link the city centre to the Millbay waterfront quarter. The Royal William Yard has been redeveloped as an award-winning Grade I listed ex-naval victualling yard. The architectural quality of building refurbishment, including residential, retail, commercial and public space, is high. There is rhetoric on the need for a more sustainable transport system for the city, but few projects emerge to support sustainable travel behaviours. For Royal William Yard, the public transport linkage to and from the city centre is poor, with only limited bus connections. Car parking on site is available, with over 400 public spaces at inexpensive rates (for example, £2 for 2 hours). There are no direct cycle connections into the city centre or around outer urban areas, hence most journeys to and from the new development are made by car. Similarly, the regeneration of Devonport, carried out with nearly £50 million of funding from the Labour government's New Deal for Communities Programme (from 2004 onwards), has improved some of the housing stock, replacing some of the postwar housing that was in a poor condition. But, again, there is little consideration given to the transport strategy; there are still poor public transport and cycling connections into the city centre, and no significant transport projects were included in the regeneration programme.

Two edge-of-centre new residential neighbourhoods further illustrate the huge problems in moving towards more sustainable travel behaviours. The discursive meaning continues in that it is difficult to move beyond motorisation – even the contemporary so-called 'sustainable neighbourhoods' are car-dependent in design with very little public transport or cycling facilities. They are incredibly depressing to view, representing the state of transport and urban planning in the UK. The first is Sherford, a new planned neighbourhood, with an envisaged 5,500 new homes, a resident population of 12,000, 7,000 jobs, four schools and local retailing, including a local high street. The masterplan states that this will be an exemplar sustainable community: 'a modern town in a traditional setting' (Sherford Building Futures, 2018). Construction started in 2015 and is expected to last for 20 years. The masterplan was developed with the Prince's Foundation (2004) as urban designers and the housing is designed in a traditional (pastiche) classical style, similar to Poundbury in Dorset. The neighbourhood includes new open space, play areas and community facilities. Internal movements can presumably be carried out by walking and cycling, yet there is much street and open space given to car parking, either off or on road, and it is free to use, even when on the street. Public transport connections are very poor, with only a limited bus service provided, involving a 45-minute journey into Plymouth city centre. The main connection into Plymouth is via the A38 highway. Most trips into the city will be by car and use this route. There is also inexpensive car parking in Plymouth. The location of the new development is too remote from Plymouth and public transport connections have not been delivered, alongside very weak travel demand management measures – hence journeys will overwhelmingly be undertaken by the car. All of this is not evident on reading the masterplan and transport assessment documentation – the development is presented as high-quality, progressive sustainable urban development, with projected high non-car mode shares. Hence, there is a problem in the procedures used to project transport movements – they massively underestimate the likely traffic implications.

Second, and even worse, is the new neighbourhood at Plymstock Quarry. This is located within the urban area of Plymouth and is a smaller-scale development, with 1,500 new homes planned on the old quarry site. The new development is similarly promoted as a high-quality sustainable neighbourhood. Yet, the housing is of poor quality with little open space. There is a segregated cycle route along the disused railway into the city centre but, again, the site has poor public transport connections. Most streets in the new neighbourhood are filled with on-road parked cars and even the supposed central open space is partly used as a car park. Again, the vast majority of journeys to and from the neighbourhood will be by car. The new primary school is surrounded by a large car park, presumably envisaging that many children will be delivered by car. The quality of these new neighbourhoods does not compare well to the better international practice being developed in cities such as Freiburg, Copenhagen, Malmö, Utrecht and others. Here, transport and urban-planning practice seems to deliver housing of higher quality and better environmental performance, and the open space and community facilities help people to walk and cycle locally. There are usually good connections into the neighbouring urban centres by public transport, such as by rail or tram, and less car parking provided in residential areas.

Hence, across Plymouth, transport planning remains hugely problematic, with too much traffic capacity and priority across the city, too little reallocation of streetspace to other means of travel, plentiful and inexpensive car parking, little high-quality public transport or bus priority, and few high-quality, segregated cycling facilities being considered or implemented. There is no effective plan for sustainable transport in the city centre or outer suburban areas and, therefore, travel behaviours will remain car-dependent. Unfortunately, this is not a problem unique to Plymouth, it is found across the majority of cities and urban areas in the UK, beyond London and the core cities, and indeed wider still in many cities internationally.

# Part III
# Truth

## The struggle for truth

Foucault often searches for 'truth' in his work, but with a particular focus on seeking to reinstate a form of truth that has been marginalised (Foucault, 1961, 1975, O'Farrell, 2005). He argues that there are historically and culturally specific framings of truth, including a related set of apparatus, rules and institutions. Truth is therefore the subject of struggles for power, and even many supposed 'facts' are an interpretation and have been deemed, at some stage and via certain actors, as 'acceptable knowledge' (Foucault, 1967). Sometimes Foucault is perceived as denying objectivity and pursuing relativism, that is, proposing that there is no single objective truth. However, this misunderstands the importance of the framing of truth. In terms of transport planning, this can involve the examination of differing forms of mobility, including motorisation-dominant transport systems and wider systems using public transport, walking and cycling as more dominant modes. These forms of truth can be analysed in varied cities, where differing transport systems have been developed and travel behaviours are more or less sustainable in environmental and social terms. Society is often supportive of the existing formation of the transport system, assuming that this representation of space and travel is the way 'it should be', hence wider views on potential uses of space can be difficult to uncover.

A number of discursive concepts are important to consider whilst discussing elements of truth. Contemporary society is based on notions of the 'normal', used as a principle of coercion, and often implemented through conformity to codes, laws or social practices. Normalisation, hence, becomes an instrument of power (Foucault, 1975, 1997) and is often used to provide the narrative for a particular mode of transport. In many contexts, car usage is presented as the usual, expected or aspirational form of travel, and other means of travel are given a lesser priority. For example, the bus is routinely presented as only suitable for lower-income passengers, who happen not to have a car; or the cyclist or pedestrian are presented as 'deviant', avoiding traffic rules on the highway, and are overlooked in much of conventional highway planning. Access to employment, education, retail, health and leisure facilities is often structured, through spatial planning (including its poor application), to require use of the car, hence reinforcing the need to use the car as the only possible or sensible option. Wider problems are related to accessibility levels, such as poverty, already unevenly distributed by income, gender and ethnicity, but reinforced by transport provision.

The Greek word *techne* is used to denote the practical rationality governed by a conscious aim (Foucault, 1988a), that is, understood as the technical skill, technique, process and know-how used to further a position. This is distinctive to episteme (which represents the structures underlying the production of knowledge) and praxis (the action). In transport planning, this has very specific and important applications as the knowledge and set of techniques and processes used within the transport discipline, including the dominance of quantitative analysis, transport modelling and forms of project appraisal. There may be a relatively fixed dichotomy between the understood and thinkable, and other possibilities that are deemed as unthinkable in particular contexts, for example, certain projects are considered as inappropriate or unimplementable. Regimes of truth are the mechanisms producing the discourses which function as true in a particular time and place (Foucault, 1975).

These issues are examined using five case studies, each demonstrating distinct and separate 'truths', developed in a particular context and then often replicated in wider cities. The case studies are as follows: Oxford (traffic demand management in the city centre); Freiburg (tram-based urban extensions alongside high-quality cycle networks); Singapore (traffic demand management, mass rapid transit and satellite new towns); Bogotá (bus rapid transit and cycling improving access for lower-income neighbourhoods); and Houten (cycling as the primary basis for travel in a new town). Each of these progressive discourses has been developed over decades and become important for practice internationally, with other cities following the lead, or at least aspiring to this form of transport system. These cities demonstrate that transport systems can be implemented without dependence on the private car. They also illustrate that social equity impacts are important for transport planning, for example, there needs to be consideration of who is using the different elements of the transport system and why. Foucault was interested in exposing the exercises of power leading to social injustice and the story of motorisation is an example of this. Impossibly, yet incredibly, the private car has become central to travel behaviours in many contexts. The case studies demonstrate that very different transport pathways can be produced in cities, reflecting differing perspectives on what can be planned and implemented, that is, the cities have developed their own regimes of truth on transport, with consequential impacts on travel behaviours for their local populations.

# 5
# Oxford

## Early rejection of highway building

Oxford faces many of the transport problems found in smaller cities and urban areas in the UK, but accentuated due to the higher travel demands associated with being a well-known university and tourist city (Jones, 1989). The current resident population is 165,000 (Oxford City Council, 2025) and growing a little year on year. The medieval centre has many historic buildings, including those associated with the University of Oxford, and hence there is restricted space for new transport infrastructure and the use of streetspace has many competing demands.

Oxfordshire County Council is responsible for transport in the city of Oxford and, over recent decades, has applied a progressive approach to transport planning, at least relative to practice for smaller urban areas in the UK. There has been a consistent application of traffic demand management (TDM) and a series of projects that attempt to promote sustainable mobility. The transport strategy introduced TDM in response to rising car use and traffic congestion, originating from the 1950s onwards. Car usage has remained relatively low in Oxford, particularly for trips to the city centre, where traffic volumes are 25 per cent lower today than in the 1990s (Oxfordshire County Council, 2019). The car mode share for journey to work trips is at 32 per cent in Oxford (Office for National Statistics, 2011), although use of the car is more prevalent across the Oxfordshire region, with a higher car mode share, at 54 per cent, compared to public transport (train and bus) at 10 per cent, walking at 11 per cent and cycling at 7 per cent.

Highway building plans were considered in Oxford in the 1960s, but rejected due to public controversy over potential impacts on the historic city centre. A highway expansion plan, originating from T. Lawrence

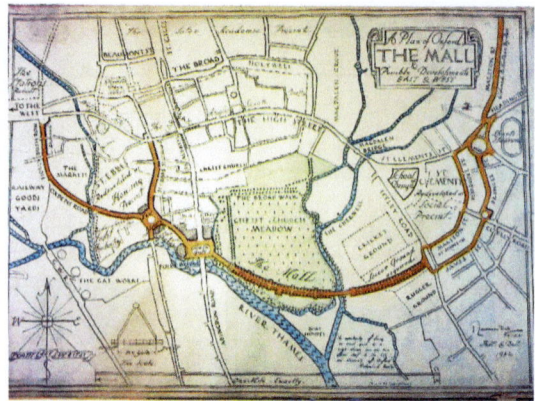

**Figure 5.1** The southern relief road proposals for Oxford.
Source: Godfrey, 2021, Digital Bodleian, https://creativecommons.org/licenses/by-nc/4.0/.

Dale in 1942, aimed to provide an inner southern relief road, linking Iffley Road to Abingdon Road and crossing Christ Church meadow (Godfrey, 2021) (Figure 5.1). The project and different route alignments were discussed at public inquiries in 1963 and 1965, but were hugely controversial, not least with Christ Church College, the wider academic community and supposedly some leading national politicians. The relief road was officially rejected in 1973.

## A balanced strategy

Instead, a revised approach was conceived to reduce traffic in the city centre. This was framed as a 'balanced' transport strategy (Oxford City Council, 1973), using the positive connotation to suggest credibility. The strategy included traffic restriction measures in the city centre and improvements to bus services from the surrounding region into the city. This included the world's purported first park and ride interchange on the edge of the urban area, where cars could park and the bus be taken into the city centre. Pedestrian provision was improved in the central retail area as space was taken away from traffic and car parking. The parking supply was, in effect, moved to the edge of the city: 20 per cent of the central area parking supply was provided in the park and ride car parks, or nearly 50 per cent if only public spaces are counted. The system has been well used and is perceived as being successful, despite some concerns over the cost of operation and the generation of additional car usage to access the park and ride sites (Parkhurst, 1995). There are now five peripheral park and ride sites with five thousand parking spaces. The system has led to a switch from car to bus trips, and less vehicles in the city centre, hence facilitating the space for the transformation of the central area streets and surrounding environment. But this latter and critical beneficial impact is, of course, difficult to measure and depends on wider policy measures. The park and ride model has been particularly influential and replicated in many cities in the UK. The concept of the balanced transport strategy has also been applied in many other cities and regions, understood (and normalised) as the 'sensible' way to produce a transport strategy, using a balance of TDM and investments in public transport. This was reformulated later as the avoid-shift-improve framework (ASIF) (Dalkmann and Brannigan, 2007, Hickman et al., 2011), which added in vehicle technology improvements to the strategy, and is now well used in many contexts through sustainable urban mobility plans. There have also been integrated transport strategies (focusing on integration between the modes). These framings have been useful to facilitate greater political and public support for transport strategies, that is, it is difficult to argue against balance and integration. However, often, the level of 'balance' or 'integration' is insufficient, and the funding given to public transport, walking and cycling, and the extent of TDM, could be much more significant, leading to higher non-car mode shares.

In Oxford's city centre, commuter trips and through traffic were gradually restricted, incrementally over decades, with access for shopping trips given priority. The two main streets in the city centre, Queen Street and Cornmarket, were closed to through traffic in 1970–3, with the exception of buses and vehicle deliveries. Alongside, public transport was encouraged using a bus-based strategy (Jones, 1989). However, there were wider political factors at play. Buses were deregulated in 1986 in areas beyond London, hence making coordination of services difficult across the city and region. Bus services mostly use the High Street for access through the city centre and hence this is a very busy route, despite having many historic buildings. There are many vehicles and a constrained streetspace, resulting in an unattractive route for pedestrians and cyclists. Cornmarket was fully pedestrianised in 1999 and further city centre streets were pedestrianised in 2009–12, including closure of the High Street for private vehicles, using a bus gate.

Full pedestrianisation was planned for Queen Street alongside the opening of the £500 million refurbished Westgate shopping centre in 2017 (Figure 5.2). But, Queen Street was a project determined by the UK Department for Transport and the Secretary of State for Transport; the bus companies lobbied for continued access, and it was agreed to allow 30 buses an hour to continue using the street. The buses now awkwardly travel in one direction at very slow speeds of under 10 mph, alongside a large volume of pedestrians. This was a classic British compromise, which works neither for pedestrians nor even bus passengers. Broad Street, similarly, has had various plans for streetspace reallocation and pedestrianisation over recent years, including an experimental use as the Broad Street 'meadow' during the Covid-19 restrictions. The meadow temporarily included a grass lawn, wildflowers and outdoor seating – and was fantastically popular with visitors. However, this has since been removed, and, even in a street surrounded by such a collection of historical buildings, it seems difficult to permanently remove the traffic and car parking. The public consultation on the project gave overwhelming support and it is hoped a permanent pedestrianisation project will be completed at some stage.

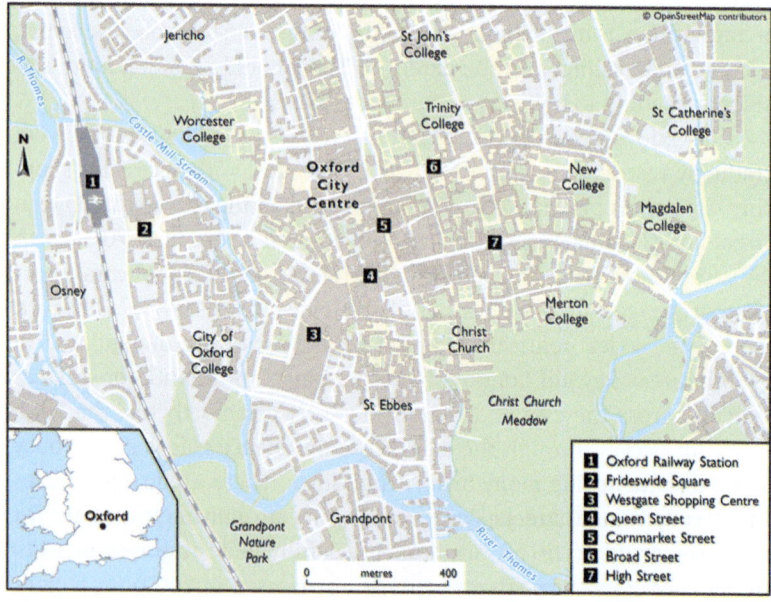

**Figure 5.2**  Transport and streetscape interventions in Oxford.

**Figure 5.3** Oxford and the bicycle. The bicycle is synonymous with the city, or at least the university. But surprisingly the transport networks are very undeveloped – cyclists do not have many segregated, high-quality facilities.

**Figure 5.4** Broad Street. The historic built environment is unique and of high quality. But the streetscapes detract and the transport infrastructure and traffic paraphernalia do not support the urban fabric. Even Broad Street is still accessible to the private car and there are no dedicated cycling routes.

**Figure 5.5** Christ Church meadow, nearly lost to the near-catastrophic EVENT of the proposed southern relief road in the 1970s.

**Figure 5.6** Bus gate on High Street. The DISCURSIVE FORMATION of the 'balanced' transport strategy has led to traffic restrictions in the city centre and bus priority. But transport planning needs to be much more progressive to achieve higher non-car mode shares and environmental and social equity goals. The balanced approach inspired the avoid-shift-improve framework, which has been applied across many wider contexts.

**Figure 5.7** Queen Street and semi-pedestrianisation, not yet fully-pedestrianised despite large pedestrian volumes.

**Figure 5.8** Frideswide Square and shared space. A shared space design has been used in Frideswide Square, as an innovative TECHNE (technical approach) to balance walking, cycling, bus and private vehicle movements on the street.

**Figure 5.9** Oxford railway station. But, still, the cycling and public transport experience is of poor quality relative to what is possible.

**Figure 5.10** Woodstock and car dominance. Car dependence is much greater beyond the city centre of Oxford. The small market towns often have a high-quality built environment, but are overrun with cars. The REGIME of transport is still providing high levels of motorisation in Oxfordshire.

## Recent streetspace reallocation

Wider streetscape space reallocation projects are an important part of the transport strategy. For example, Frideswide Square was redesigned in 2015, giving a shared space for traffic, pedestrians and cyclists. This was inspired by shared space designs originating in the Netherlands, from traffic engineers such as Hans Monderman, and involved a very different techne, or technical approach, to dealing with street design. The design seeks to slow traffic speeds, minimise demarcations such as traffic signing and white lines, allow more space for pedestrians and cyclists, and rely on interaction and eye contact, between pedestrians, cyclists and motorists, to manage the use of space (Hamilton-Baillie, 2008). This is distinct to earlier traffic-planning approaches aimed at segregating the different modes. 37,000 vehicles, 20,000 pedestrians and 2,500 cyclists use the junction at Frideswide Square each day, hence this remains a well-used route. The project cost £5.8m and significantly improves the route into Oxford from the railway station. A similar design has been implemented, at a smaller scale, at the junction of Broad Street and Parks Road. There are continuing plans to refurbish Oxford railway station, alongside the new Oxford Parkway station (opened in 2015), which gives new services to Marylebone station in London, and potentially to Milton Keynes and Cambridge as part of the Oxford–Cambridge Arc regional planning strategy.

To further tackle air quality, a low emission bus zone was introduced in Oxford in 2014, which involved the change of the bus fleet to Euro V compliance for nitrogen oxide. This has been extended to Euro VI compliance over the period 2015–7. Further plans are for black cab taxis to be 'zero emission capable' by 2025, that is, vehicles should emit less than 50 grams/kilometre of $CO_2$ and be able to drive over 100 kilometres with electric power. A zero emission zone is being introduced, involving a £2 charge for vehicles entering the central area and emitting more than 75 grams/kilometre of $CO_2$ from 2021, rising to a £4 charge in 2025 (Kraftl, 2021). In addition, a workplace parking levy of £4–600 per year per vehicle is planned for parking in the Eastern Arc of Oxford, covering areas such as the Oxford business park, a lower density, sprawling and car-dependent industrial area. This is part of the Connected Oxford initiative (Oxfordshire County Council, 2019) and will allow car usage to be tackled in the eastern part of Oxford and also generate local transport funding, which can be spent on transport infrastructure.

Transport planning is hence relatively progressive in Oxford. But, there are major difficulties in developing an effective cycling network or public transport system due to the political structure for transport planning in the UK. For example, there is a high level of cycling in the city centre, but there are very few segregated cycling routes and cycle parking provision is very poor. There has been too little funding available for high-quality cycling facilities over decades. Bus deregulation means there is little control over bus routes, frequencies or costing, as the operators remain fragmented. The regime of truth, in this case of bus deregulation, proffered since the Thatcher government in 1986, has been difficult to move away from. It is evidently clear that privatisation does not lead to more efficient and higher-quality bus service delivery, yet this was the narrative that was produced and applied. Only in 2023 were some metropolitan authorities moving away from this biopower, such as in Manchester and Liverpool, despite decades of poor bus service delivery. It is difficult, in the current context in Oxford, to upgrade the more popular bus routes into enhanced public transport provision, such as a tramway. This could very effectively serve the city centre, linking to the railway station, Cowley, and perhaps Headington and Witney. A central Oxford tunnel would assist in removing buses and vehicles from the High Street, and has been mentioned in the latest transport strategy (Oxfordshire County Council, 2015). But this type of perceived 'radical' measure remains virtually impossible to fund and implement in the UK context and remains largely undiscussed, stranded with the label of 'politically impossible' to deliver. There are examples of similar projects being implemented in mainland Europe, such as in Karlsruhe. Hence the undiscussed narrative remains particular to the Oxford context and the specific discourse that has been developed.

If this status is considered in terms of motility, there is a particular field of possibilities that has been produced, with an associated level of appropriation leading to actual travel behaviours. There is a so-called balanced transport strategy, developed over decades. This leads to some use of the bus, but with fragmented operators and poor bus services; some cycling, despite few segregated routes; and limited walking facilities and public space. Transport provision still does not support the historic built environment, adversely impacting the city. There is continued reliance on car usage in the surrounding county of Oxfordshire. These narratives have all become fairly fixed and normalised, so that any significant changes are difficult to implement or even discuss. The disciplinary power of motorisation is exercised, despite and through its invisibility. Funding for transport is highly centralised, with approval for larger projects having to be sought from the UK government, via the Department for Transport, using a specified set of project appraisal procedures (Department for Transport, 2024). This is a peculiar way of developing transport planning, meaning that the county authority has to gain approval and bid for funds from the national level to implement large projects. The techne (the set of technical skills and procedures) and the episteme (the structures underlying knowledge) have been developed over decades, and the approved process for transport projects favours particular modes and specific locations. The system is particularly unsuited to implement public transport, streetspace reallocation or walking and cycling projects. The strategy and transport-planning procedures are not focused on achieving wide-ranging public policy goals (Hickman, 2019). Rail and bus services remain limited and expensive and housing is so unaffordable that lower- and even medium-income groups cannot afford to live in Oxford or some of the surrounding county towns. Transport-planning practice in the UK is hence bounded by the discursive practice and formation, within the given political and societal context, and does not allow more significant progress to be made towards sustainable transport.

# 6
# Freiburg

## Environmental consciousness from the 1970s

Freiburg, located in the southwest of Germany, has been known as a leading city for environmental-based urban planning since the 1970s. As Hall (2014, 251) describes 'it is the city that has done it all'. Freiburg is a small and affluent university city, with a population of 230,000. Employment is mostly associated with tourism, the city government and the university, which is one of Germany's oldest, founded by the Austrians and dating back to 1457.

The environmental movement has developed over decades in Germany and has had a large influence on public policy. The planned construction of a nuclear power station in the early 1970s, just 30 kilometres from Freiburg, was pivotal in developing the local environmental consciousness in the city. Civil disobedience over the building of the power station led to the plans being dropped in 1975 and a wide environmental coalition was formed, including left-leaning students, local political parties, university staff and local civic leaders, and extending to church leaders and the surrounding farming community. Concerns over the Chernobyl disaster in 1986 and the impact of acid rain on the neighbouring Black Forest led Freiburg to strengthen their environmental policies for the city. A tradition of citizen participation in local politics emerged, which became important to environmental, transport and planning policies in the city, and continues today. Hence, a critical element of the techne (the practical rationality) is the level of awareness and participation of the public – and environmentalism becomes the regime of truth, producing the discourse for transport and urban planning that is widely accepted in society.

## Traffic restriction in the city centre

Freiburg's historic city centre was almost completely destroyed in WWII, but rebuilt in the historic style, contrary to most cities in Germany, which were rebuilt with a modern architectural style. The city authority approved its Reconstruction Plan in 1948, specifying that the city should be rebuilt in its historic and compact form – and this is what you now see on visiting the city centre. The older land use and transport plans from 1955, 1963 and 1969 favoured growth and urban expansion, and were based on facilitating growth in the use of the private car. Some of the old streetcar lines were even abandoned, services cut and car parking allowed in the city centre squares. Motorisation grew rapidly from 28 vehicles/1,000 population in 1950 to 248 vehicles/1,000 population in 1970, which was higher than the German average at the time. However, the later modernisation plans were never fully approved by the city and public opinion shifted from developing the city around principles of urban dispersal and car-based mobility, towards a stronger environmental perspective and the compact city ideal. The city was, therefore, protected and developed as a compact city, including by developing public transport, walking and cycling networks. The first Cycle Network Plan was developed in 1970, and even by 1972 there were nearly 30 kilometres of cycle paths in the city. The streetcar network was preserved from 1972 and expanded as a tram network (Stadtbahn) from 1978 onwards, with additional lines built and an environmental ticket (a monthly pass for unlimited use of trams and buses) introduced in 1984. The spatial plans and transport strategies developed the environmental vision over decades, for example, the Land Use Plan in 1981 focused development around public transport corridors, while the Transport Plan in 1989 aimed to reduce car usage and prioritise public transport, walking and cycling (Buehler and Pucher, 2011).

Freiburg provides an early example of restricting access to private cars in the city centre – a 'truth' that became accepted in the city and a forerunner for many cities that have used traffic demand management (TDM) and pedestrianisation to improve their city centres (Figure 6.1). In 1973, the Altstadt (old city centre) was converted into a pedestrian-only zone and was the largest such zone in Germany at the time. The ring road to the south of the city centre was closed to private cars and reserved for trams, cyclists and pedestrians, giving a vehicle-free street in front of the Stadttheater. Through-traffic was moved westwards towards the station. Hence, streetspace was freed up

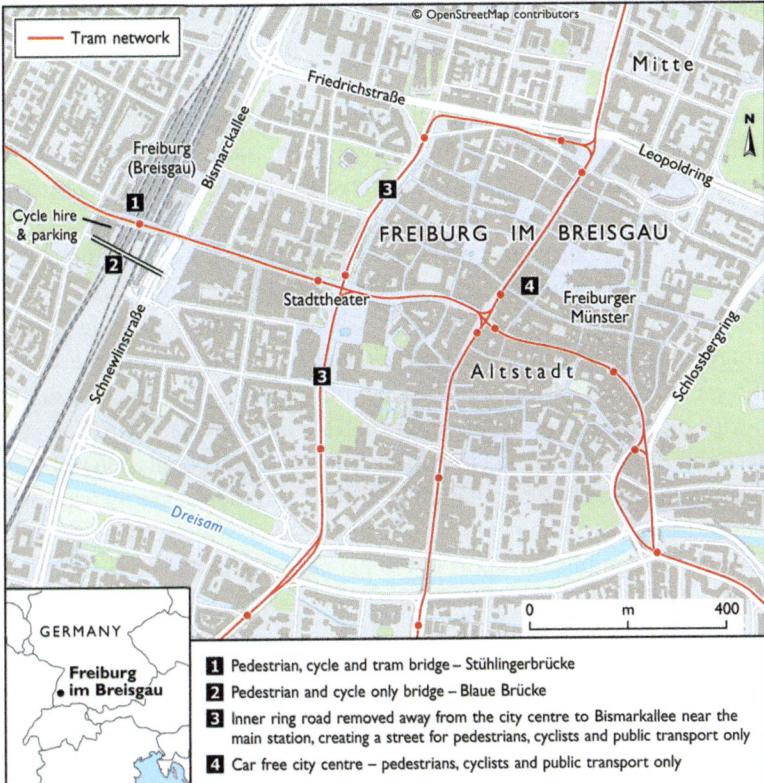

**Figure 6.1** TDM in the city centre of Freiburg.

for pedestrians and cyclists in the city centre. Car parking was removed from historic squares in the central area, such as Cathedral Square, which is home to Freiburger Münster (the grand gothic cathedral) and now hosts a popular market. Traffic was removed from bridges over the central railway station in the 1990s; one bridge was preserved for trams (Stühlingerbrücke) and another for cyclists and pedestrians only. The latter iron bridge (Blaue Brücke) is used by more than 10,000 cyclists per day and is a popular place to sit and dwell for students and tourists. All residential areas have a speed limit of 30 kilometres per hour, at most, and car clubs are available for occasional vehicular usage. Hence, over decades, a very successful traffic management strategy has been developed, which effectively reduces traffic in the city centre.

The 'city of short distances' concept (Hamiduddin, 2015) was gradually developed and refined, with a tram network and extensive cycle network serving the city. Today, there are over 30 kilometres of tram lines, connecting to extensive bus services, a regional railway network and 400 kilometres of cycle paths. A RegioCard has replaced the previous public transport card and is available for around €50 per month, allowing unlimited use of the urban transit and regional routes. Indeed, in 2022, there was a promotion allowing unlimited public transport travel across Germany for €9. Many of the cycle paths in Freiburg provide segregated routes through open spaces and between neighbourhoods and there are 9,000 bicycle parking spaces in the city, including cycle parking, cycle hire and a café at the central station. Over the last three decades, integrated land use and transport policies have helped to increase the share of trips by walking, bicycle and public transport, and reduced the share of trips by private car. The motorisation rate has held steady in Freiburg since the 1990s, decreasing from 422 vehicles/1,000 population in 1990 to 419 vehicles/1,000 population in 2006. Car mode share by trips has reduced (from 38 to 32 per cent) over 1982–2007, whilst cycling (15–27 per cent) and public transport (11–8 per cent) have increased (Buehler and Pucher, 2011).

**Figure 6.2** Freiburg city centre is attractive for pedestrians and the car has been removed. Environmental-led planning has developed over decades and become the REGIME OF TRUTH (the mechanism producing the discourse that is accepted).

**Figure 6.3** Blaue Brücke and active mobility. The bridge provides a walk and cycle connection over the railway and direct 'filtered' connection into the city centre.

**Figure 6.4** Tramway and the Vauban neighbourhood, built to low-energy standards with direct tram connections and segregated cycle pathways into the city centre.

**Figure 6.5** Community space in Vauban. The lack of streetspace given to car access and car parking becomes NORMALISED as residents realise that the spaces can be used for children's play and other community uses.

**Figure 6.6** Shared garden areas. The design of space helps to encourage social interaction.

**Figure 6.7** Rieselfeld neighbourhood. There are central areas for social interaction and the selling of local produce.

**Figure 6.8** Car parking at the rear of buildings in Rieselfeld. Space for cars is removed from the front of houses and provided in rear basement or perimeter car parking – so that walking, cycling and public transport become the most convenient modes and streetspace is freed up for community uses.

**Figure 6.9** Cycle facilities. There is an excellent segregated cycle network across the city and secure cycle parking in residential areas and in the city centre.

**Figure 6.10** Compact urban planning. The DISCURSIVE FORMATION of compact urban planning is applied and has led Vauban and Rieselfeld to become well-known examples of 'sustainable neighbourhoods' – yet this seems difficult to reproduce in wider contexts.

## Suburban neighbourhood extensions

A further impressive element of the city's environmental planning has been the development of new neighbourhoods. Now over 20 years old, two suburban residential neighbourhoods – Vauban and Rieselfeld – have been built with employment, retail, education and recreational facilities. The buildings are built to low-energy standards, heated by a combined heat and power station (burning woodchips) and solar panels. The neighbourhoods are linked to the city centre by extended tram lines and cycle routes, with strictly limited car access and car parking (Figure 6.11).

Vauban, in particular, has been well-known for decades as an exemplar of residential neighbourhood extensions. It was built on the grounds of an old French military barracks, and is named after the seventeenth-century French military leader who built fortifications in Freiburg when

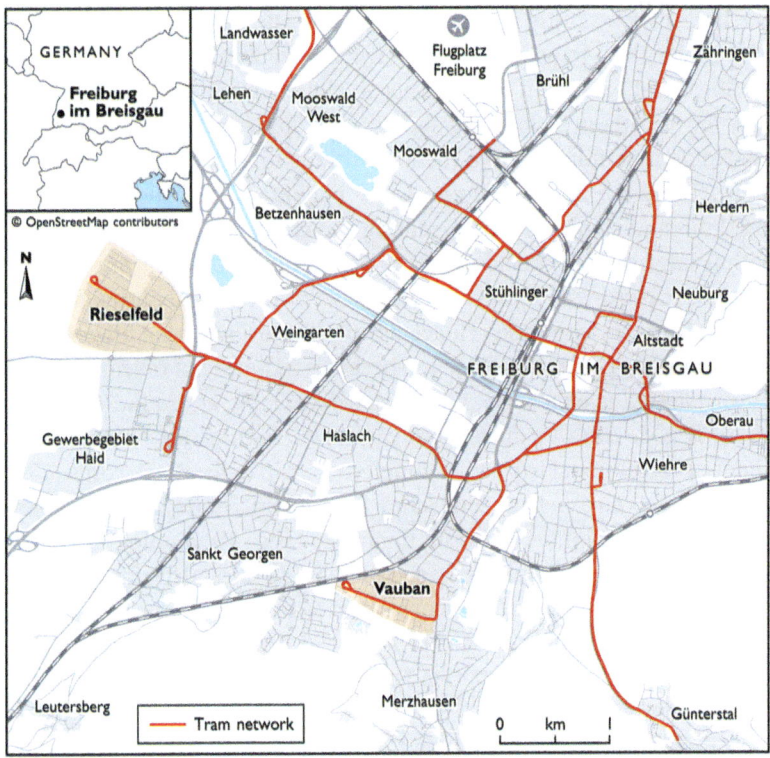

**Figure 6.11** Vauban and Rieselfeld neighbourhood extensions in Freiburg.

under French rule. The neighbourhood provides 2,200 houses and flats for a population of around 5,000. It was designed by the city authority with the input of the residential community through the Forum Vauban, and built mostly between 1993 and 2006. A tram route travels from the entrance of the neighbourhood through the central avenue, surrounded by four–five storey residential and mixed-use developments. There is a hotel, schools, shops, small-scale employment and other community facilities – and many everyday trips are possible by walking or cycling. A range of housing types are spread around the development, including for low incomes. There are Baugruppen (groupbuild), where interested purchasers come together and collectively fund their multi-housing units, and communal housing. Vauban was initially very popular with young families, due to the layout and generous communal open space, though the population has aged over the years. The community open space remains, and this is the outstanding feature that is evident if car ownership and car parking can be reduced.

The main streets are restricted to vehicle speeds of 30 kilometres per hour or less; most residential streets are designed as home zones, with space given to pedestrians and cyclists, and only occasional access is allowed for cars, with speed limits at 7 kilometres per hour. There are very few on-street car parking spaces, with space in front and between the residential units given to the local residents to design as communal areas. There are many different uses – play areas for children, including slides and climbing frames; seating and communal barbecues; all set within landscaping, including many of the old trees from pre-barrack times. Car parking is removed to the edge of the neighbourhood, with two multi-storey car parks. Hence use of the car involves a longer walk than accessing the tram or bus, and becomes a choice not taken. The cost of car parking is expensive, with a space available at around €18,000. There is a car club for infrequent car usage, such as for buying bulky goods or recreational trips, for example, a ski trip to the Black Forest. The tram provides an excellent connection into the city centre, with a journey time of just 15 minutes and tram frequency every 10 minutes. The journey by car takes much longer, hence there is intentional 'traffic filtering', which makes car usage more inconvenient than other modes. As a result, car ownership has been kept very low (150 cars per 1,000 population in 2006, relative to 420/1,000 in Freiburg and 560/1,000 in Germany) (Buehler and Pucher, 2011).

Rieselfeld provides a second new neighbourhood development, built mostly from 1994–2010 on a former sewage farm. It is a higher density development, with residential, retail, educational, employment and other community facilities. There are around 4,000 homes and 10,000 residents (Hamiduddin, 2015). There is more car parking provision under and at the rear of residential units, relative to Vauban, but again the provision is low at around 1.5 spaces per unit, and this has maintained low car ownership and usage. Again, a high level of community space has been maintained, with much landscaping and open space. The wider urban planning in Freiburg has helped keep private car usage low: edge-of-town retailing, for example, is prohibited due to the negative impact on city centre retailing and the dependence on the car for visiting these types of retail developments.

## Environmental-led, participatory planning

The truth produced in Freiburg is one of environmental-led urban planning, including low-energy buildings and high levels of public transport, walking and cycling. The praxis (action) is to consider how most effectively to implement the policy approach, with citizen participation as a central element in this. The process reflects the theories of communicative action, as developed by Habermas (1984), with residents and other key actors exchanging ideas on what is possible and helping to develop programmes and pathways of action. Freiburg has a particular population that favours sustainable travel behaviours; this is a city dominated by the university and public sector, with high levels of education and income alongside widespread pro-environmental beliefs. Environmental principles have been important in shaping the planning of the city, including new neighbourhoods and the transport system. The problematisation of public policy and reshaping of practice has been led by the Green Party, which regularly gains a high share of voting in city elections. There has also been a remarkable coherence and consistency in urban planning and political decision-making over decades. The city government has been effective and innovative, with Wulf Daseking acting as Head of Planning in the city authority for nearly three decades, between 1984 and 2012. Alongside, the same mayor often retained power for decades, such as Rolf Böhme, a socialist from the Social Democratic Party (1982–2002) and Dieter Salomon from the Green Party (2002–18). Perhaps less progress has been made in recent years, as housing becomes more unaffordable and new neighbourhood extensions have not been produced. But, certainly, there is an exemplary basis for environmental lifestyles that can be further developed. The more difficult task is to deliver these types of urban neighbourhoods, transport networks and travel behaviours in wider political and social contexts, where there is a population with less support for environmental goals. But, revisiting Vauban and Rieselfeld, we can see that the high quality of development has been maintained; these are still excellent places to visit and live, decades after having been conceived and implemented. The landscaping has matured, the housing build quality has endured, and the public transport and cycling networks are still well used. This remains an exemplar for high-quality, environmentally sustainable urban living, but one that seems difficult to reproduce where the levels of environmental awareness and support are not evident.

# 7
# Singapore

## Urban expansion amid space constraints

Singapore has developed a progressive mobility strategy over decades, including innovative measures. The discursive formation of integrated transport planning and urban planning includes restrictions on private vehicle ownership and use, an extensive Metro system and high-density development in the satellite new towns (Barter, 2013, Diao, 2018, Barter, 2019b). Singapore is a city-state with a population of 5.9 million (2023) and the island has an area of 734 square kilometres, hence there are severe constraints on space for new development and population growth. The city-state is also wealthy, derived from its history as a trading port, and now as a centre for the financial sector and tourism in Asia. Average income levels are at S$62,000 (Singapore Government, 2023b). But, there is also social inequity with large disparity in incomes, as exemplified by a Gini coefficient of 0.44 in 2021 (Singapore Government, 2023a). Politically, the city-state is also unique, ruled as a parliamentary representative democratic republic, but with one-party rule by the People's Action Party (PAP) since self-governance in 1959 and independence in 1965. This gives a particular episteme (unconscious structure) for policy making, which facilitates an effective implementation of interventions, including participation and debate, but with limits and controls on contestation.

## Developing an integrated planning and transport strategy

From the 1970s, increasing population and income levels led to rising housing demands and traffic growth. Space constraints in the city-state acted as an important driver for spatial planning and, particularly, transport planning. Singapore's governance approach, or apparatus, is technocratic, with a well-resourced, in-house public sector, enabling long-term and consistent plans to be pursued. The Urban Redevelopment Authority (URA) is responsible for urban planning; the Housing & Development Board (HDB) for public housing; and the Land Transport Authority (LTA) for transport planning. There is cross-sectoral working, hence integrated plans can be made that work towards different policy goals over the long term. The latest mode share data is from the 2016 household survey, with private vehicle at 29 per cent of trips, bus at 29 per cent, mass rapid transit (MRT) at 24 per cent, walk and cycle at 14 per cent (but cycling almost negligible) and taxi/private hire at 4 per cent (Land Transport Authority, 2018). Vehicle usage has hence been held at low levels for years, with high patronage for public transport, but there is much potential for increasing walking and cycling.

The strategic approach to urban planning and transport planning was initially developed in the Concept Plan (1971), with the city centre, satellite new towns and industrial areas located in a ring around the coastline and central water body, separated by open spaces and linked by the expressways and public transport. The plan encouraged urban growth and high-density satellite new towns, alongside a comprehensive network of expressways, traffic demand management, the development of MRT, and a new airport at Changi, all of which have been implemented to varying degrees. The Concept Plan was updated in 1991, 2001 and 2011, with a related White Paper in 1996, and a Land Transport Masterplan in 2008 and 2013, furthering the initial approach that had been set. The bus system was the only form of public transport in the early years, with a requirement for fares to match operational costs, and different operational models were used. The MRT system was introduced from 1987 and provided a much-improved public transport service. Alongside, an extensive network of expressways was planned and implemented, whilst walking and cycling facilities were largely overlooked. Gradually, the city centre has grown in size and over 20 satellite new towns have been built. These accommodate much of the population of Singapore (Field, 1992), for example, Toa Payoh (developed from the 1960s onwards), Serangoon (1980s) and Tengah (2020s). The MRT system has developed into an extensive network. The current Long Term Plan (Urban Redevelopment Authority, 2022) outlines the strategic approach to urban planning and transport, with a continued focus on polycentric growth to reduce movements into a monocentric city centre. The Masterplan (Urban Redevelopment Authority, 2019) gives a more detailed spatial strategy for different neighbourhoods. The Land Transport Masterplan (Land Transport Authority, 2019) provides the transport strategy, drawing on the 20-minute city concept and applying this to the island context. The aim is to develop a 45-minute city across the island, where access is possible to employment and wider activities across the city; and 20-minute towns, with access at the neighbourhood level to everyday activities such as schools and local retailing.

Traffic demand management, including controlling private vehicle ownership and use, has been the most innovative element to the transport strategy over decades. From 1975, the road pricing system was introduced, known as the area licensing scheme (ALS), and was initially paper-based. This was replaced by electronic road pricing (ERP) in 1998. Transponders are fitted on vehicles and a charge is applied on an area basis, as gantries are passed on the main highways in the city centre. The current charges vary by the level of congestion, hence spatially and by time of day, but can be around S$10 dollars for each gantry passed in the peak. The system is not distance-based or emissions-based, hence the size and emissions of vehicles do not affect the charges. Vehicle purchase tax was also initially used to make private car purchase more expensive, subsequently replaced by a vehicle quota system from 1990. This was particularly innovative and involves direct management of vehicle numbers. Individuals can apply for a certificate of vehicle entitlement (COE) and a charge is set relative to demand. Only slow growth in vehicle numbers was allowed in the early years and then a cap set at one million vehicles from 2008, which still exists today. This seems an obvious approach for transport planners seeking to move away from individual car usage, but, of course, it is rarely used in other cities. The COE has varied in cost over time, but is currently expensive, at around S$100,000 for a 10-year licence. Ownership of a typical small car, such as a Toyota Prius C, with the vehicle costing S$20,000, might involve a COE of S$85,000, an additional registration fee of 110 per cent (S$22,000) and customs excise import duty of 30 per cent (S$6,000), hence an actual overall cost of S$133,000 (£78,000). The COE is linked to the vehicle, therefore, a second-hand car is sold with the remaining years left of the COE. COE also provides a significant income stream, estimated at S$6.9 billion in 2016 (Diao, 2018). Alongside further income from ERP and motor vehicle tax, this can be used to fund wider infrastructure development. The management of vehicle numbers and charging for road usage have become a normalised part of transport planning in Singapore, and seem broadly accepted as effective means to manage traffic volumes. Vehicle taxation and road pricing are, however, flawed policy instruments in distributional terms. They are regressive in facilitating higher-income drivers to pay the charges and to use the relatively free-flowing roads, whilst lower or even middle-income groups are dissuaded from car ownership. Again, this seems an accepted discourse, reflecting a regime of truth; that the road system is prioritised for higher-income car drivers. But, the income streams are significant and this helps fund public transport and wider infrastructure for wider users. Hence, the system can also be viewed as a tax on car drivers for hypothecation elsewhere. Singapore has around 100 vehicles per 1,000 population (in 2012). This overlooks that vehicles are shared within and between households and vehicle kilometres travelled (VKT) are high at around 18,000 kilometres per annum (Barter, 2013, Diao, 2018). Yet, the result is that car ownership is held at low levels and there are relatively low numbers of vehicles on the roads. The city has much lower levels of congestion than other cities in southeast Asia and the wider continent.

Car parking policy in Singapore is more conventionally implemented, seeking to meet demand with supply (Barter, 2019a). Hence, there have been few attempts to use car parking as a traffic management tool. There is mostly ample resident and visitor parking provision in the HDB residential estates with low usage charges and wider on-street residential car parking is seldom charged or managed. City centre parking is provided in multi-storey car parks, surface car parks or on-street (either URA managed or private), and is more expensive at S$2–5 per hour or S$300–600 per month, but low relative to European city standards. There is often free car parking for access to parks and open spaces. Minimum car parking standards for new developments have been used since the 1960s and have only recently (2018) been complemented with maximum car parking standards acting as an upper limit.

Alongside the management of traffic, the MRT system has continually been extended (Figure 7.1). MRT provides an extensive, integrated and easy-to-use network, covering most of the island and giving good access to the city centre and the satellite town centres. Usage is more equitable across population groups, as the MRT has good spatial coverage and fares are relatively low; a journey is charged at S$2, using a two-hour period as a single journey. Most stations are accessible, with escalators and lifts available for people with disabilities. New public transport projects are being planned, such as the Thomson-East Coast Line (TEL), Jurong Region Line (JRL) and the Cross Island Line (CRL) (Land Transport Authority, 2019). The bus and taxi fleets are slowly transitioning to use low emission vehicles, with 'cleaner' diesel-electric hybrid vehicles planned by 2040; however, electric vehicles could be much more speedily introduced.

The current transport-planning approach is described as 'car-lite' (Land Transport Authority, 2019), yet there are tensions in the approaches taken. There is a dense network of highway provision, with three- and four-lane highways throughout much of the urban area; hence a contradiction in providing so much highway space alongside other traffic demand management measures (Barter, 2019b). There have been many expressway projects, giving grade-separated highway access between the urban centres. These include the 12-kilometre Kallang-Paya Lebar Expressway (completed in 2008) and the 5-kilometre underground Marina Coastal Expressway (completed in 2013), which is partly in tunnel under the sea. There is landscaping of highway corridors, which improves the visual appearance, but there is still much land take and severance from the highway network. Traffic congestion is relatively low, even in the peak periods, and vehicle movement is free flowing. But there are few attempts to reduce highway capacity for vehicles, even in the historic central areas, and streetspace reallocation is rarely attempted. Alongside traffic ownership restriction, public transport and urban planning, Singapore has provided a very generous highway network for a relatively select cohort of users. The high cost of vehicle purchase and

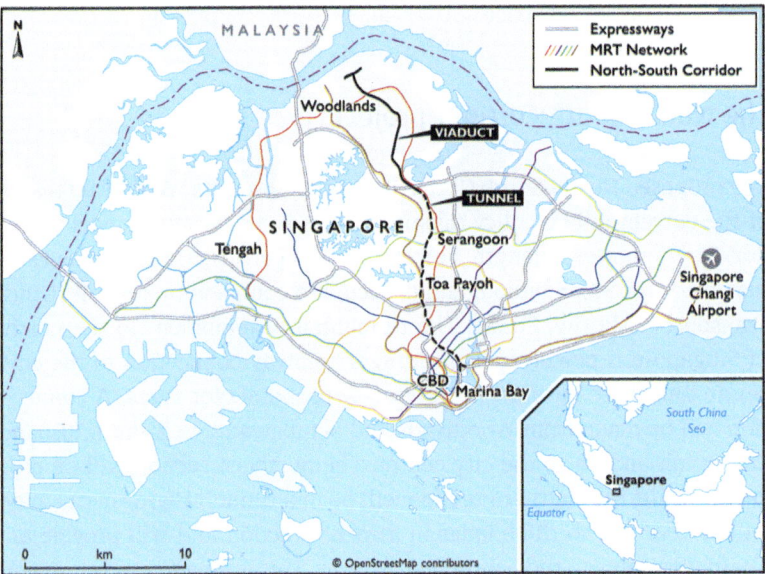

**Figure 7.1** Singapore MRT and the proposed North–South Corridor.

restriction of the consumer good leads to a high status for car ownership, which again is problematic.

Active travel has been given far less priority since the 1970s. There is a significant level of walking in terms of trips made, yet footways can be narrow and uncomfortable to use. Routes are often circuitous via pedestrian crossings over the highways, waiting times for pedestrian crossings are lengthy and crossing times are short for slower pedestrians, such as the elderly. Indeed, 'jaywalking' (walking across the road at a so-called 'undesignated' place) is famously prohibited in Singapore and subject to fines, yet rarely enforced. There are some adaptations to the climate for pedestrian facilities, such as the daily heavy rain, and many footpaths have covered linkways, particularly to and from MRT stations. There is surprisingly little high-quality pedestrian space in the city centre, including in the bay area and historic areas. Alongside, there is almost no provision for cycling in the city centre, with few segregated cycle pathways and only a few emerging leisure routes, usually discontinuous at key locations. There are more cycling facilities in the satellite towns, with cycling used for park connectors and as the 'first mile' route into the MRT system. There are plans for a growing cycle network, with 1,000 kilometres of routes planned by 2024 (Land Transport Authority, 2019), yet the quality of this needs much improvement from current routes and to move beyond the leisure and first mile functions. There are, as yet, few serious attempts to reallocate space away from the private car to give more pedestrian or cycling space.

## The North–South Corridor project

The proposed North–South Corridor project presents a window on many of the current public policy dilemmas. This is a new multimodal corridor, 21 kilometres in length, currently under construction, with an estimated cost of S$7.5 billion, and expected to complete in 2027–9 (Land Transport Authority, 2023). The residential development to the north of Singapore is perceived as needing a highway connection to the city centre and there have been expressway proposals for years. A connection will be made from Woodlands and Sembawang, to Yishun, Bishan, Toa Payoh and on to the city centre. The northern section will be elevated, whilst the southern section will be in a tunnel. Highway capacity will involve two to three lanes in a each direction and will provide an alternative route to the Central Expressway. This is seen as the 'last major highway capacity project' on the island, but of course this narrative has been used to justify many highway projects that are subsequently superseded by further highway projects. The project has been repurposed to improve bus, pedestrian and cycle facilities at surface level as well as provide highway capacity. The visual character of the corridor at surface level will therefore be improved relative to its current usage as a highway. The main debate here is the choice to continue to increase highway capacity alongside the improved bus, walk and cycle connections. Presumably a Metro- or bus-only connection could have been developed and the improved highway capacity has not been provided. However, this possibility remains undiscussed in the promotional material for the project – the options given are very tightly bounded and do not reflect the potential to invest only in sustainable urban mobility solutions. The narrative, of course, is very positive, suggesting that 'the streets will be returned to the people', 'spaces will be transformed' to 'create the future together'. This reflects the historical discourse of transport planning in Singapore and a continued trajectory of providing for improved public transport options, but, alongside, also improving highway capacity. A much stronger application of sustainable urban mobility is possible, removing space for private vehicles, or at least not adding capacity, and instead providing high-quality public transport, walking and cycling.

**Figure 7.2** Singapore Marina Bay. The governance structure for the city-state leads to a particular EPISTEME that facilitates innovative policy making and effective implementation, but also with limited opportunity for contestation.

**Figure 7.3** Electronic road pricing as the TECHNE. The integrated approach of urban planning, MRT and TDM have effectively reduced traffic volumes – including a limit of one million vehicles in Singapore.

**Figure 7.4** The extensive MRT system has been developed into an extensive network, providing integrated connections across the city.

**Figure 7.5** Toa Payoh new town. The new towns include high-density residential development, mixed-use commercial centres and connections into the MRT. They provide for high-quality suburban living with social interaction in the community spaces.

**Figure 7.6** Footway provision. There are some good pedestrian facilities, but this can be extended, including through streetspace reallocation.

**Figure 7.7** Chinatown and the shophouses. Some historic areas are well conserved and provide distinctive features within the urban fabric.

**Figure 7.8** The North–South Corridor project provides an opportunity to improve surface-level walking and cycling facilities – but is also providing increased highway capacity as an elevated route or in tunnel.

**Figure 7.9** Highway provision. There is extensive highway space in Singapore, and this has been NORMALISED within transport-planning practice, despite the problematic environmental and distributional impacts.

## Wider possibilities?

Singapore set an early pace for progressive transport planning in Asia, from the early 1970s onwards, with traffic demand management, an extensive MRT and integration with urban planning. The control of car ownership, through the COE, is the most innovative measure applied in Singapore. This is probably only implementable in a political structure (episteme) as found in the city-state, though similar approaches have been used in China. Singapore has been very influenced by understanding transport planning through the discipline of transport economics – so that pricing for travel becomes the central intervention tool. This has been useful to an extent, but it also overlooks the distributional issues and the importance of infrastructure provision. The inertia in highway planning continues and there are proposed projects, such as the North–South Corridor, which will deliver large increases in traffic capacity. Alongside, the extensive MRT system gives excellent public transport coverage across the island-state alongside the bus system. The new towns are well planned, giving local retail and commercial centres, which allow everyday activities to be reached and some level of self-containment within the neighbourhoods. In the coming years, there is much potential for improving the walking and cycling networks. This can only really be done by reducing the streetspace given to the car and reallocating space towards walking and cycling. This would also improve the liveability of Singapore, with the severance of the highways reduced and active travel becoming more possible and space returned for new public spaces. There are planned new developments that can provide low car living, such as the planned development of reclaimed land at Marina Bay South. This could even become a model for an eco-neighbourhood with low car ownership in Singapore and wider Asia. But, this would mark a significant change in approach for Singapore. There are other significant projects on the horizon, such as the proposed high-speed rail (HSR) connection to Malaysia. This can dramatically improve connections to Malaysia, and onwards to Thailand and potentially further to China. This would give Singapore much wider access into the labour market of Malaysia and would feasibly change travel behaviours quite dramatically, as housing and labour costs are cheaper in Malaysia. Responding to climate change in Singapore demands a fuller range of interventions that can more significantly reduce transport $CO_2$ emissions. At the moment, there is insufficient transition to sustainable travel behaviours. Hence, there needs to be much greater discussion and debate on appropriate policy options, including how active travel might be more encouraged.

# 8
# Bogotá

## Public transport and improved accessibility

Bogotá, the capital of Colombia, is well-known for its progressive sustainable mobility initiatives. The urban area is home to over eight million residents and three million more in the metropolitan region beyond, with successive waves of migrants fleeing poverty, civil war and violence in the rural areas (Kimmelman, 2023). The city is the third highest capital in the world, set on a plateau in the Andes at an average altitude of 2,640 metres above sea level. The eastern hills and Bogotá River constrain the shape and growth of the city into a linear form. A key feature, similar to many South American cities, is the marked spatial segregation by income. The lower-income neighbourhoods are found on the edge of the city, mostly to the south, and higher-income neighbourhoods are located in the central areas, mostly to the north. Neighbourhoods are classified into six social stratification groups based on dwelling characteristics and the urban environment. Taxes are higher in the higher-income areas and these cross-subsidise utilities and facilities in the lower-income areas. The existence of 'red-lining' around the neighbourhoods has some drawbacks, with the poor location reputation leading to difficulties for residents when seeking employment and general status (Guzman et al., 2018).

The historic transport network, running from the 1880s to the 1950s, included horse-powered and then electric streetcar trams and urban rail. These were removed and replaced with bus, trolley bus and a growing number of private cars. Motorisation was encouraged, following a masterplan by Le Corbusier and US-based consultants, Town Planning Associates, in 1950. Urban highways were built to accommodate motorisation, though the trolleybuses operated until 1991. Alongside, a chaotic and privately operated microbus system grew from the 1930s onwards. But, the informal system became problematic as it struggled to carry increasing volumes of passengers around the city. The microbuses were perceived as hugely inefficient, with thousands of independent minibus operators competing for patronage. Service routes were unclear and of poor quality, there was over and under supply in different corridors, poor integration between bus providers, low-quality vehicles and traffic congestion and air pollution. Corruption was an additional issue, with a lack of public control over bus services and mistrust of the private operators. In many cities, this has been an unwritten reason for more regulated public transport services to be pursued through city-operated bus systems.

By the 1990s, Bogotá had become a city with severe urban problems, including financial crisis, social deprivation, inequity and crime. Mayors Antanas Mockus (1995–7 and 2001–3) and Enrique Peñalosa (1998–2000 and 2016–9) sought to transform the city, increasing public infrastructure investment, gradually leading to improved tax revenues and reduced crime. Investments were made to urban infrastructure, including in water, electricity and education. Transport connections were improved, particularly for the low-income communities on the edge of the city, together with improved public spaces. An extensive bus rapid transit (BRT) system was built, known as TransMilenio – and this became central to the discursive formation of transport planning in the city. Earlier plans for a network of elevated highways and a Metro system, following the recommendations of consultancy studies from JICA and Ingetec S.A-Bechtel-Systra (1997), were rejected. Instead, Enrique Peñalosa improved the quality and integration of public transport, with strategic planning by the city authority and private bus operation (Montezuma, 2005). The first TransMilenio line opened in Bogotá in 2000 with subsequent lines developed (Figure 8.1). This was supported by traffic restriction, such as license plate recognition, and an extensive network of cycle pathways, with 270 kilometres of routes built by 2003. TransMilenio became the leading example of BRT in South America, and was often perceived as the 'template' BRT system for other cities to follow. TransMilenio was itself modelled on an earlier BRT system developed in Curitiba, Brazil, built from 1974, and known as the Rede Integrada de Transporte (RIT).

TransMilenio used a hybrid public–private model similar to that used for the Curitiba RIT: the city authority was responsible for the planning, management, construction and maintenance of the infrastructure system, whilst the vehicles were run by private sector operators under concession contracts. This is the financial model advocated

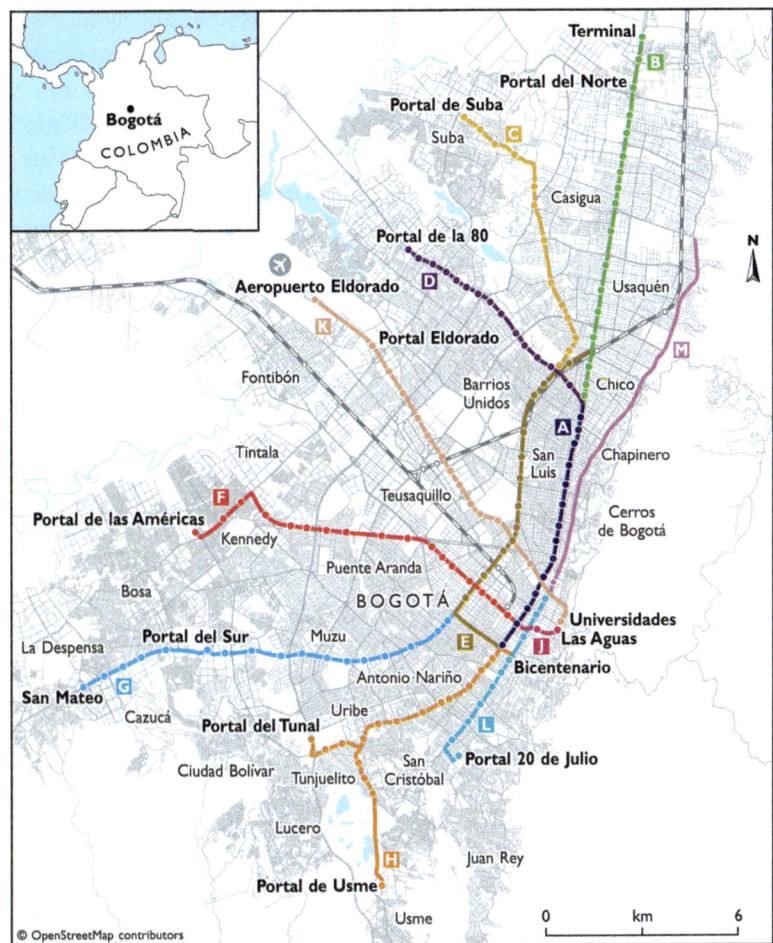

**Figure 8.1**  The TransMilenio network in Bogotá.

for transport infrastructure projects, over decades, by actors such as the World Bank (Gwilliam, 2002). The Colombian central government provided 70 per cent of the funding and the remainder was from the city of Bogotá (Hidalgo et al., 2013). Infrastructure and wider cost estimations vary, but are around $428 million for phase one, with 41 kilometres; and $454 million for phase two, with 63 kilometres. Infrastructure cost is around $4 million per kilometre, 10 or 20 times cheaper than a Metro equivalent (Gilbert, 2008). This was a key argument for developing BRT – patronage can get near to that of a Metro system, but the cost of development is much less, though of course journey speeds are slower.

The early years of TransMilenio were perceived as hugely successful. An extensive network was developed, carrying large volumes of passengers at high speeds on segregated lanes. The system is viewed as a high capacity BRT, carrying 14,000 passengers on the first day (Hidalgo et al., 2013). Implementation of the project was seen as a significant success in a difficult context, with difficult social and mobility problems and recurrent public policy failures. The World Bank has subsequently part-funded many similar systems, including following the public–private operating model which accords with their neoliberal views on public policy and aid funding (Gilbert, 2008). The distinctive red-coloured articulated buses have a capacity of 160 passengers per vehicle; and latterly large bi-articulated buses were utilised, with increased capacity for 270 passengers per vehicle. The network has grown to include 12 lines, covering 115 kilometres across the city, with over 1,600 buses on the trunk lines and nearly 800 on the feeder lines. There are regular, express and super express buses, carrying 2.4 million passengers per weekday and 680 million annually (2018). 45,000 passengers per hour per direction can be carried on the most heavily loaded sections (Hidalgo et al., 2013).

There are many innovative design features to TransMilenio (Hidalgo and Carrigan, 2010, Hidalgo et al., 2013), including vehicles use bus-only lanes, dedicated bus stations, pre-boarding payment using a smartcard at a turnstile, level boarding direct onto the bus, bus and station doors open simultaneously on boarding, and there is centralised system control. The main corridors are often two lanes in each direction (hence four lanes across the highway corridor), allowing express services and overtaking at bus stations. Passengers typically reach the raised-floor stations in the centre of the road via pedestrian bridges. A fixed fare is used, hence the longer journeys from the poorer suburbs are subsidised by the shorter journeys from the higher-income central areas – an innovative fare strategy driven by social equity concerns. Feeder services serve the outer neighbourhoods, with green-coloured vehicles bringing users into the main corridors via the terminal 'portal' stations. The feeder services usually run without dedicated lanes, but are free, and passengers are only charged when entering the main corridors. Hence the system is socially progressive, supporting usage by the lower-income groups. Secure cycle parking is provided at many of the terminal stations, often with large capacity and cycle hire and repair facilities. The TransMilenio logo is well-known and used on the buses, stations and maps, making the system very distinctive. The BRT system became an iconic symbol for the renewal of the city. The regime of truth produced in Bogotá reflects a regulated BRT system that could move many passengers around the city, much more effectively than the previous microbuses or investment in highway capacity for the use of the car. BRT was perceived as more effective than Metro investment due to the reduced implementation cost per kilometre. There is normalisation of using public investment, through building an extensive system and improving connectivity across the city, and high patronage levels follow. The network is particularly focused on social equity goals, with access improved to and from lower-income neighbourhoods into the city centre's activities.

However, over time, TransMilenio became overused, suffered from capacity problems and lacked appropriate investment. Though still delivering large volumes of passengers, there were some challenges in operation. The network has not been extended as far as originally envisaged and there are problems including crowding, long journeys and waiting times. There is bus congestion on key routes and in the city centre, alongside traffic congestion (Moscoso et al., 2019). Fares are relatively expensive for poorer cohorts, higher than conventional bus services, at 2,000 pesos per journey ($0.60). This level of fare can take up a significant share of a household budget. Bus stations and even road surfaces have deteriorated. There are financial problems with some bus operators, including bankruptcy. User and non-user ratings have declined from 2007 onwards, mainly due to the crowding and safety issues whilst travelling, such as theft and sexual harassment and assault (Guzman et al., 2018). There are sometimes operator strikes and public demonstrations concerning journey conditions. Critically, the vehicle fleet used diesel fuel for two decades, leading to much air pollution on the city's streets. Only recently, in 2022, have clean vehicles been mandated through procurement processes. But, much progress has been made, and the electric bus fleet is now the largest for a city outside China. The large BRT corridors also dominate the streetscape, meaning that the streets are very intrusive on the surrounding neighbourhoods. Space for pedestrians and cyclists is often constrained. The Bogotá Metro has been proposed for decades as a complementary network to BRT. But, this has been very slow to be implemented, and the first line is only currently in construction (planned to be operational in 2028). It is estimated that the Metro will carry 72,000 passengers per hour per direction and one million passengers per day, hence adding much-required capacity to the city public transport system.

There are further unwritten narratives associated with TransMilenio. The episteme (the structures underlying the production of knowledge) framing public policy interventions means that the system and any potential extensions are viewed through a neoliberal transport economics lens. The expectation is that public transport is completely or mostly profitable, even in day-to-day operational terms. Paget-Seekins (2015) queries this financial model, involving complex and lengthy contracts for fleet purchase, infrastructure provision and operation. The private operators are paid by the city authority for passenger and vehicle kilometres travelled, with some subsidy for fares and security. The city authority, through TransMilenio, hence provides the infrastructure for private operators to run at profit, leading to higher ticket prices for users and giving less potential to reinvest funds into enhanced services. Operation is usually by large private bus companies and the smaller bus operators are not involved. Over time, this financial model severely constrains extending the network, improving the quality of waiting environments and the surrounding public spaces, including building new lines to more peripheral neighbourhoods. Some cities, such as Medellín, have moved away from these procedures and provide services directly through the city authority. This avoids many of these operating problems, but of course involves higher upfront costs for the city.

## Cycling and low-income neighbourhoods

A further component of the success story of sustainable mobility in Bogotá has been in cycling. The cycle network has been gradually extended and improved across the city, including a focus on connecting low-income neighbourhoods. This helps to improve access to the different neighbourhoods and city centre, including for employment, education and leisure activities. An example is the Alameda El Porvenir cycle route, which links two low-income neighbourhoods in the southwest of the city. The cycle route was built before the development took place in the neighbourhoods, hence becoming the 'spine' to the area, shaping how it grew. The cycle routes also integrate with the BRT interchange stations, providing 'last mile' connectivity to the bus network, including secure spaces to park the bikes. Cycling has grown quickly in usage; in 2015, mode share for cycling was at 5 per cent of trips, with TransMilenio at 15 per cent, other bus at 29 per cent, pedestrian at 30 per cent and private car at 11 per cent. Later surveys indicate cycling mode share has risen to over 10 per cent of trips and more than one million trips are made by bikes in Bogotá every day (Secretaria Distrital de Movilidad Bogotá, 2019). Most recently, cycling levels have levelled off as motorcycle usage has grown – as a response to increased congestion on the BRT system and highways.

**Figure 8.2** Highway provision. The potential rise in car ownership and usage has been averted with a revised focus on investment in public transport and cycling.

**Figure 8.3** The TransMilenio network has become the DISCURSIVE FORMATION of transport planning in Bogotá. The system facilitates the movement of large volumes of passengers across the city, but has become crowded and needs continued enhancement.

**Figure 8.4** City centre public spaces. The central areas are accessible by public transport and the public spaces are well designed – but, again, continued investment is required to maintain quality in provision.

**Figure 8.5** Cycle provision. Cycling facilities have been well provided, including with experimental projects that help build support amongst the public.

**Figure 8.6** Alameda El Porvenir cycle route. Some of the cycle routes provide excellent connections for the lower-income neighbourhoods. The EXPERIENCE of travel is much improved for disadvantaged groups.

**Figure 8.7** The Ciclovía and active travel. The EVENT has been important in showing that cycling is possible for many people, helping develop a cycling culture and facilitating social interaction across the city.

**Figure 8.8** Cable car access. Journeys in the city have been enhanced for many disadvantaged groups, including by income and spatially. The provision of transport facilities directly for deprived neighbourhoods is NORMALISED. This is a key element of transport planning in Bogotá.

## The Ciclovía

Bogotá's well-known Ciclovía ('cycleway') has been central to the development of cycling facilities and a cycling culture in the city, alongside the extensive formal cycle network. The history of the event stretches back to 1974 when local activists started to organise cycle rides. In 1976, the Ciclovía was formalised by the city authorities and promoted as a city-led event. The Ciclovía involves the closure of key roads to traffic on Sundays and national holidays, and gives the streetspace to pedestrians, cyclists, runners and skaters. Roads are closed from 7am to 2pm, using a simple method, such as cones and tape, blocking entry for vehicles. The road closures have gradually been extended over time, now providing routes of over 121 kilometres in length across different parts of the city (Figure 8.9).

The Ciclovía involves more than a street closure or a pedestrian and cycling event. Alongside the widespread walking and cycling, further activities are held in parks, neighbourhoods and the city centre, such as music, rumba, yoga and aerobics. The event has become a celebration of city life, held every week, and is extensive and hugely popular. The event is regularly used by around two million people, which is 30 per cent of

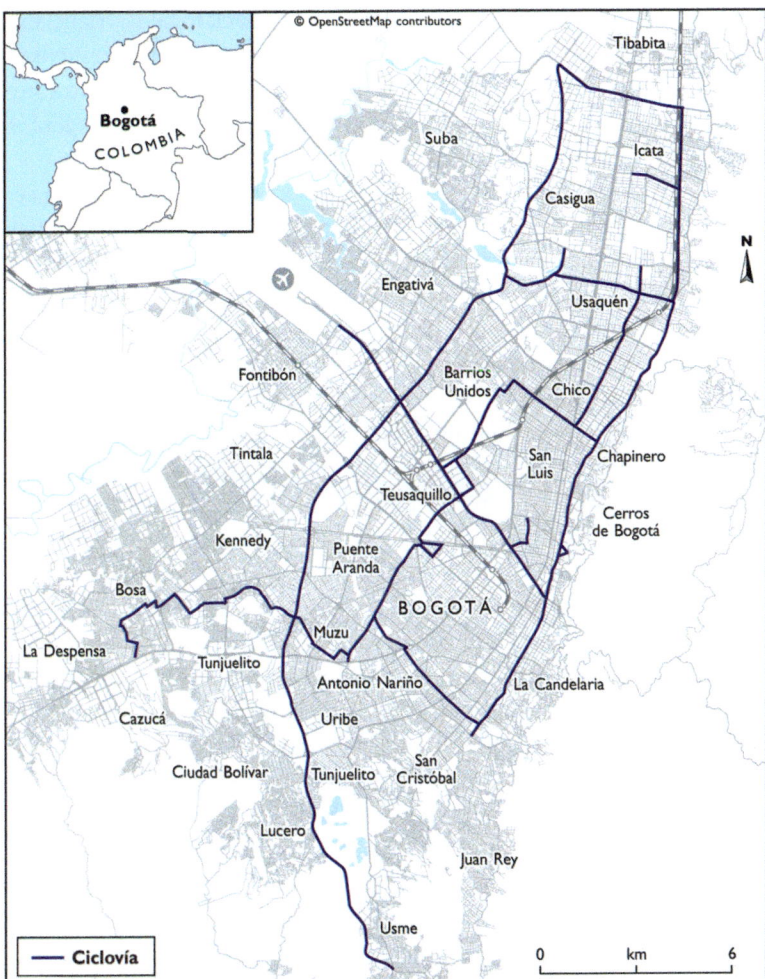

**Figure 8.9** Bogotá's Ciclovía.

the city's population. The Ciclovía helps transform areas that are perceived as unsafe into welcoming and friendly places, with people travelling through high-, middle- and low-income neighbourhoods, mixing freely, gaining physical activity and interacting. Similar events are now held in many parts of South America, North America and Asia, for example, in Buenos Aires, Medellín, Lima, Mexico City, Denver, Miami and Portland. Wider events are held internationally under different names, such as Car Free Days, Critical Mass Rides, Reclaim the Streets and Open Streets. These events are successful and popular when undertaken, demonstrating the impacts of giving streetspace back to pedestrians and cyclists. The takeup in the UK and wider continental Europe contexts is interestingly less than that in South America, perhaps demonstrating that the levels of political grassroots activism in cities such as Bogotá is strong and can lead to public policy change in the city authority.

Many lessons can be derived from the Bogotá transport experience, but perhaps four are most evident. First, the range of mobility possibilities for the lower-income suburban neighbourhoods was massively increased by TransMilenio and the cycling networks, and these led to significant positive social equity impacts. But, second, as the city grows, the public transport networks have to be continually enhanced. Although public transport accessibility is relatively good across Bogotá, the outer areas still suffer from lengthy and uncomfortable journeys, such as from the informal settlements in the south. Feeder bus and cable car connections were built to some of the low-income informal settlements, but not all. Average travel times remain high, and are increasing over time. Hence, further investment in TransMilenio is required as mobility demands grow, alongside the new Bogotá Metro which will give much further capacity for travel. Third, the development of TransMilenio and cycling in Bogotá provides an example of targeting transport investment towards disadvantaged groups. Typically, this is not done in transport planning, which conventionally overlooks important distributional issues. Usually it is the higher-income groups that benefit from transport investments, as these use the new infrastructure or are least affected by adverse impacts such as severance or air pollution. The justification from Enrique Peñalosa (2024) was that, in giving cyclists their own space, he 'democratised' the city by giving a cyclist the same right to move as a person in an expensive car. Transport investments should not only be for the advantaged groups, but for all in society. This can be a lesson in normalisation taken from Bogotá – that distributional issues are critical and transport investment can play an important role in improving social equity, spatially and across income groups. But, finally, the epistemic framing of transport planning needs careful consideration, and the structure of neoliberal economics often constrains transport project development. Public infrastructure is required to support urban growth, and often this is much more effectively planned and implemented by the city authorities themselves, rather than losing money through interaction with private operators who seek to take profit out of the system. The narrative of neoliberalism needs to be carefully evaluated in transport, in terms of outcome, and reshaped.

# 9
# Houten

## VINEX planning

A particular form of transport system has been conceived and implemented in the Netherlands in many cities and urban areas. This is a transport system based around cycle infrastructure and use of the bicycle as the primary means of travel for short trips, combined with public transport for longer trips (Pucher and Buehler, 2008, Hickman et al., 2025). The infrastructure is critical, but there are wider place-based social and cultural factors that support cycling (Nello-Deakin and Nikolaeva, 2021). The Netherlands rejected the growth in motorisation, which occurred in the 1960s–70s, due to the energy crisis in the early 1970s and also the rise in traffic casualties which was very unpopular. In 1973, the pressure group 'Stop de Kindermoord' (Stop the Child Murders) was formed, campaigning against the rise of the private car and, instead, supporting cycling (Bicycle Dutch, 2011). This became influential in transitioning Dutch transport policy away from motorisation and towards cycling. Cycling was shaped as a symbol of national identity by associating the bicycle with values such as simplicity, practicality and modesty. The 'distinction through simplicity' appealed to the egalitarian social ethos found in the Netherlands (Oosterhuis, 2016).

Alongside, there has been a lengthy tradition of the apparatus of the national government using spatial planning at the regional, urban and neighbourhood scales. This sought to manage the spatial pattern of urban development, including a compact urban form and integration with the public transport system. This development pattern supported the use of public transport, walking and cycling. In the 1970s and 1980s, a strategic planning approach was developed by the Department of Housing, Planning and Environment (VROM) with a policy of concentrated decentralisation. This focused growth in designated urban centres and restricted the sprawl of smaller rural settlements. In the 1980s and 1990s, compact urban growth was furthered, with the urban housing stock renewed, out-of-town shopping centres were prohibited and new employment guided to locations well served by public transport under the 'ABC location policy' (Schwanen et al., 2004). Open space was protected in the centre of the Randstad region, with the 'Green Heart' protecting urban areas from sprawling in between the cities of Amsterdam, The Hague, Rotterdam and Utrecht.

A central element of this approach was the VINEX strategy (the supplement to the Fourth National Policy Document on Spatial Planning), published in 1993, where housing development was planned in a 10-year programme across multiple cities in a polycentric and compact form. VINEX provided locations for new development, coordinating new housing, industrial and commercial development with transport infrastructure investment. Nearly one million new homes were planned between 1995 and 2010, with new housing located in urban areas or as urban extensions and distributed across different urban centres, including in the Randstad and other regions (Galle and Modderman, 1997). The plan aimed to control suburbanisation, protect open spaces from development and improve access to public transport. The connection of VINEX locations relative to public transport was not always successful and there were shortages of highly accessible locations for new mixed-use developments. However, new development has generally been well focused on existing urban centres, or extensions to these; there are excellent rail connections between cities in the Netherlands, and cycling is very well used.

## Houten new town

Houten provides an example of new urban development and is a VINEX growth location (Hickman et al., 2025). The new town was planned in the 1960s by the lead masterplanner Robert Derks as a 'Groeikern' (new centre of growth) to accommodate the growing population in Utrecht and the surrounding region (Figure 9.1). The main design objectives were to create a liveable neighbourhood, set within open space, where residents knew each other and could visit their local centres and socially interact without having to use the car. Ten thousand new dwellings were constructed from 1979, expanding the small existing rural village of Houten, and the population reached thirty thousand by the 1990s. New neighbourhoods continue to be added and the population reached fifty thousand in 2022 (Statistics Netherlands, 2022).

**Figure 9.1** Houten new town.

The town is served by Houten station, which opened in 1982 and was refurbished in 2010. Local commuter trains connect to Utrecht, just 10 minutes' journey to the north, and Geldermalsen to the south. The second phase of development was planned in the 1990s, broadly following the design principles from the earlier development, but with different types of housing, including more contemporary designs, and a new neighbourhood centre. A new station, Houten Castellum, was opened in 2010 and this serves the southern town centre and residential areas.

Each neighbourhood in Houten is developed as a distinct community, with a particular architectural design, leading to much housing variety. The earlier garden village housing is found in the northern area, and some of this requires refurbishment. The later development is found to the south, including English, Italian and French-inspired neighbourhoods. The initial Houten village remains with older housing, and there are occasional rural farm buildings which are surrounded by the new development. The planning seeks to create an urban area of villages, each with its distinctive character, reflecting the neighbourhood unit principle used in the UK new towns (Perry, 1929).

The transport system is particularly distinctive, relative to UK new town practice and indeed wider internationally. A particular form of truth for transport is pursued, giving status to certain modes and journeys over others. Internal vehicular movements are restricted and instead there are high levels of cycling and walking. Public transport is well used with the direct rail link to Utrecht and beyond. The focus on cycling as the basis for internal movements was innovative when developed in the 1980s, particularly compared to other new towns which tended to be car-dependent. This was a response to the environmental concerns of the 1970s and perceived adverse impacts of car usage that were appreciated across the Netherlands. Traffic is allowed on the ring road surrounding the neighbourhood, with access given only to particular neighbourhood areas, and not between these. To drive to another neighbourhood area, residents have to revert back to the ring road and then access the local streets from the particular local entrance point. Hence, cycling and walking journeys are purposively made much quicker within the neighbourhood and these are the dominant modes. This is an early example of 'traffic filtering', which has been applied in other countries in recent years. There is an internal network of low-speed streets (with a 30 kilometre per hour limit), including traffic calming at entrance locations from the ring road and non-linear street alignments which slow traffic.

The design of the cycle network is guided by the techne, in other words, the technical skills, techniques and processes applied. In this case, the techniques of cycle planning are guided by the CROW design manual for bicycle traffic (CROW, 2016). The extensive cycle and pedestrian networks provide for most trips, with 130 kilometres of cycle routes in Houten, mostly coloured red and segregated from vehicles and pedestrians. Cycle mode share is at 28 per cent, walking at 27 per cent, public transport at 11 per cent and private car at 34 per cent of trips (Institute for Transportation and Development Policy (ITDP, 2010)). There are examples of cycle streets, where the car is expected to share the street with cycles, hence speeds are low; again, this is an innovation in practice, used for later shared space street designs. Subway or bridge connections are provided across the ring road and at the internal middle road interchange, where there is a segregated, multi-level road and cycle roundabout. Open space is provided throughout, with a linear park to the north which crosses east to west, and a city wall park to the south. There are lakes, streams and canals; open space is hence central to all of the development. Travel within and between the collection of neighbourhoods is safe by cycle and walking, even for children. Schools are located adjacent to the parkland areas and accessed by cycle paths to encourage active travel. Many schoolchildren, even in the junior schools, cycle to school without their parents. Small electric vehicles are provided for people with disabilities. Houten rail station is located at the northern town centre, using an elevated railway line which allows cycle parking and pedestrian access at ground level. The station was refurbished as a Fietstransferium (cycle transfer station) with extensive cycle parking, cycle hire and repair. OV-fiet (public transport cycle hire) is available as well as more conventional cycle hire. There is a mix of retail and commercial development in the town centre, though office development has been overprovided and is sometimes converted into residential use. The southern centre is smaller but also has the second railway station and a mix of retail and commercial developments.

## Conceived, perceived and lived space

Analysis of the cycling experience in Houten (Lu, 2022, Hickman et al., 2025) illustrates the initial aspirations for the design and the subsequent support that has developed for cycling amongst residents. The urban and transport planners conceived the space to prioritise cycling for internal travel, with segregated cycle routes and traffic removed from neighbourhoods:

> The vision was to make it possible for everyone to have a car, but not to use a car for daily transport. So you can have a car, and you can use it if you want, but it's not a necessity. We wanted everyone, especially children, to be able to cycle safely from their homes, to the schools, to other places, to their friends, to the sports, wherever you go on a daily basis. (Interview E1, Transport Planner)

> What we said in the beginning, it's faster to go from here to here with your bike than with the car. So why should you take the car if you're faster with your bike? (Interview E4, Urban Planner)

> The most important thing is to encourage people to cycle. Encouraging people to cycle is to make cycling 'safe, safe, safe'. Safety is the first thing; if it doesn't feel safe, people are not going to use it. And then it has to be smooth. It has to be comfortable and also it has to be pleasant. (Interview E1, Transport Planner)

The cycling network has been positively experienced by residents over decades and become a central element of living in Houten. The perceived and lived space is, therefore, very similar to what was envisioned as conceived space by the masterplanners. Over decades, cycling has become central to everyday life and emblematic of the Dutch way of life:

> It's so nice that you know when you're on a red bike path, that's for bikes and that's where you're safe. And it's also what I tell my kids when I go with them on the bike. When we're here on the red path, you don't have to look out for the cars. It makes it so easy. (Interview R5)

> A lot of people move to Houten from a different place. And it may not be the decisive factor for people. It's not like, oh, let's go to the most bicycle-friendly place. But it's definitely an important part of the decision. (Interview R3)

> I love cycling. It gives me joy and freedom. And also it makes me feel more fit. It's a workout two times a day when I ride to work and back. (Interview R7)

> And it's just quite, you know, in a Dutch way, quite beautiful. (Interview R4)

**Figure 9.2** Houten town centre. The DISCURSIVE FORMATION of transport planning in Houten is based around cycling. The cycle network is of exemplary quality, including extensive segregated cycle pathways giving direct access to the town centres and railway stations. Vehicular traffic is restricted for internal journeys in the neighbourhoods.

**Figure 9.3** Houten railway station. There is excellent cycle–public transport integration, with secure cycle parking and cycle hire available directly on leaving or before accessing the train (including OV-fiet, the hire cycles provided by the public transport operator).

**Figure 9.4** The segregated cycle network. The TECHNE of cycle planning includes the design standards in the CROW cycle design manual, which has become the leading international standard for cycle facility provision.

**Figure 9.5** Cycle parking in the Netherlands is highly advanced relative to other contexts, with safe and secure cycle parking, including cycle hire and repair.

**Figure 9.6** Highway access to residential areas and parking are restricted to the edge of the neighbourhood.

**Figure 9.7** Traffic filtering: street design ensures cycles and pedestrians gain quicker and more convenient journeys relative to the car, hence traffic is 'filtered'.

**Figure 9.8** The more recent neighbourhoods have more contemporary residential designs, and are surrounded by open spaces and the cycle and pedestrian networks.

**Figure 9.9** Cycling boardwalk: cycling is NORMALISED as the 'usual' means of travel for residents.

The discursive formation produced in Houten is that cycling and walking are the most convenient and practical ways to move around within the neighbourhood. This has become normalised, amongst the transport and urban planning experts, and also the residents, so that cycling is seen as the most sensible way of travelling around the neighbourhood. Using the car is rarely considered and is only perceived as an option for travel to other towns and cities beyond Houten, and usually for those more difficult to reach by rail. The cycling discourse, and the high levels of usage, are based on the excellent cycling infrastructure. But, also, there are wider factors – the elements of history, discontinuity, power, truth, ethics and subjectivity are all evident in Houten, and over time have produced the very specific transport system and travel behaviours (Hickman et al., 2025). Houten is known locally as 'paradise in paradise', offering excellence in cycling facilities and living conditions, even relative to the very impressive cycling facilities seen across the Netherlands. This has been an intentional urban and transport-planning exercise, running over decades, from the 1960s onwards – with the new development focused around excellent public transport and cycle networks and the car removed from everyday usage.

# Part IV
# Power

## Relations between actors

Power can be closely associated with and indeed influences forms of truth. As Foucault (1977, 131) advises 'truth isn't outside power, or lacking in power' – it is a production of mechanisms of power. The effects of power relations on transport systems and travel behaviour are rarely considered, but of course have enormous impacts. Transport is profoundly enmeshed in social structures, incorporating a variety of different actors, including institutions and the public. Power reflects the relationships between these individuals, actors, groups and organisations. Governments and institutions are a central element, but power can occur at any scale, including between individuals. Power is a way of regulating people's conduct, as 'a mode of action ... upon the actions of others' (Foucault, 1982b, 340); yet power only exists when it is exercised. Power can be productive in terms of producing forms of action, for example, in employment or education, such as through managing or improving performance. But, it can also be applied negatively, such as in a disciplinary and repressive manner, for example, in the hospital or institution, prison or armed forces. Activities and behaviours are managed through disciplinary techniques, for example, through the use of living and working procedures, methods of training and surveillance. Impacts of power can occur at different scales, including at the neighbourhood or individual levels, for example, in regulating and changing attitudes and behaviours. Space can be used to help implement power. In transport, this can include supporting a particular form of transport system or using the layout of streets to maintain a particular order. Resistance can also be an important feature, where the adverse effects of power or the existing order are challenged (Gordon, 1980, Faubion, 1994) or where space is used in a contrary manner (Lefebvre, 1974).

A number of concepts are related to power. Genealogy is used to describe the historical method of research, incorporating dimensions of power and political contestation. Governmentality is used to denote the institutions, techniques and procedures used to guide the conduct of others or oneself, for example, to administer the population (Foucault, 1991). Governmentality is broadened by Foucault, in his later works, as biopower, representing all forms of the government of people by others, beyond the administration of the population (Foucault, 1978a). Within the exercise of power, there are different forms of apparatus (*dispositif*). These are the institutional, physical and administrative mechanisms which maintain the exercise of power (Foucault, 1978a). Knowledge is an important element of maintaining power, and is seen as an event, rather than an innate attribute. *Connaissance* (the corpus of knowledge, such as within a discipline) is seen as distinct to *savoir* (knowledge in general, often the knowledge underlying actions or beliefs) (Foucault, 1969). Foucault was interested in how power and knowledge existed in relation to social, economic and political factors (O'Farrell, 2005). He countered the assumption that there was opposition between power and knowledge, instead arguing that knowledge was often shaped to reflect the needs of powerful actors. For example, within science, there are procedures that help systematise how knowledge and specific data are collected to monitor behaviours. As part of this, subjects are posed, concepts formulated, and propositions constructed, formalised and validated (Foucault, 1969). All of these affect the shape of transport systems, how we travel and the associated impacts.

These issues are examined with three case studies, each demonstrating different elements of power and application of knowledge associated with transport and urban planning. The transport and urban development strategies and projects are from Chongqing (the development of a modern subway network and connection into the national high-speed rail network), London (redevelopment of the King's Cross transit-orientated neighbourhood) and Rio de Janeiro (redevelopment of the Porto Maravilha area). The discussion of knowledge is seen in terms of connaissance and savoir, both of which somehow come to be, helping to frame the understanding and application of urban and transport planning in the particular contexts.

# 10
# Chongqing

## Population growth and mobility

Many cities in China have dramatically changed their transport systems and travel behaviours over recent decades. China was known as the 'kingdom of the bicycle' in the 1950s–80s, with cycling representing over 50 per cent of mode share for trips in many cities in these decades. But, cycling has since been marginalised, with less space given to cycling on the street as the demand for 'modern' cities and travel behaviours has increased. Subsequently, cycling has declined and there has been rapid growth in the use of the private car. Substantial investments have been made in highway capacity, alongside rising household incomes, increased car ownership and individual mobility. More recently, cities have invested heavily in extensive public transport systems, including Metro and national rail. Most tier one cities (the largest cities with high incomes and material consumption) and tier two cities (the next largest) have built extensive Metro systems over the last two decades, together with surface-level rail and bus networks. In addition, there are high-speed rail (HSR) connections between cities, hence most of the intra-urban and longer distance journeys can be undertaken by public transport.

## Public transport and urban expansion

Chongqing, a tier one city located in the middle/southwest of China, provides an example of this remarkable investment in public transport. The city now has an extensive public transport network, including multiple forms. There is an urban population of 16 million and a wider metropolitan region population of 30 million (2020 census), with the historic central part of the city located at the confluence of the Yangtze and Jialing rivers. There are large changes in elevation between the rivers and surrounding mountains, giving a very distinctive topography to the built environment. Chongqing is one of four municipalities under the direct administration of the central government (the others are Beijing, Shanghai and Tianjin), and hence has important elements of governmental apparatus that support urban development. This reflects its close connections to Beijing and important profile as the gateway to and catalyst for economic growth in western China. Indeed, Chongqing acted as the provisional national capital in the Second Sino-Japanese War (1937–45), with Japan capturing Nanjing, Beijing and Shanghai. Hence, the city has historical linkages to important dimensions of institutional power and resources. Chongqing has experienced strong economic growth (>10 per cent per annum) over the last two decades, with the main industries including automobile, motorcycle and electronics manufacturers. The city has perversely benefited from the resettlement of communities from the Three Gorges Dam project, as many people were displaced to adjacent urban areas, such as Chongqing, together with the wider movement of migrants from rural areas. The city is also central to the 'One Belt, One Road' initiative, which seeks to improve connectivity and economic growth across central Eurasia (Bao et al., 2019).

The extensive Chongqing Rail Transit (CRT) network includes 10 Metro lines, with a total length of nearly 500 kilometres (longer than the London Underground system). Most lines run underground, but some are elevated monorail. The Metro system has been gradually extended since the first line opened in 2004. There are many bridges over the river valleys, including the world's highest Metro-only bridge, the Caijia Rail Transit Bridge on Line 6. There are three HSR stations (Chongqing North, Shapingba and Chongqing West), another is under construction (Chongqing East), and there are plans to refurbish the old central railway station. The network is to be extended with additional Metro lines by 2035 (Figure 10.1). In addition, there are two cable cars running over the Yangtze River, a legacy from the pre-Metro days, and these are now used mostly by tourists to access the views and appreciate the historical travel over the river. Ferries also run along the Yangtze and Jialing Rivers and are mostly used by tourists. Highway construction in the city has been limited due to the mountainous topography, and there is only one orbital highway route. This has helped Chongqing to avoid the more extensive highway capacity found in other Chinese cities. A car registration plate system is in operation to further manage traffic volumes. Highways are still evident in the built environment, with associated problems of severance, noise and air pollution, but not to the extent of cities such as Beijing.

The 14th Five-Year Plan for Chongqing (Chongqing Municipal Government, 2020) outlines the current spatial development strategy and major infrastructure proposals for 2021–5. The focus is on 'people-centred', high-quality urban development, using transit-orientated design as a design principle. Neighbourhoods are being built around the public transport system, such as the Chongqing East Railway Station HSR new district. Modern architecture and higher densities are found in the city centre, including the eight towers and connected Crystal enclosed skyway at Raffles City, opening in 2019. The contemporary design by Safdie Architects also reflects the historic and imperial city gate and maritime trading on the waterfront.

The extensive public transport infrastructure is used to accommodate increased passenger movements around the city and enable urban

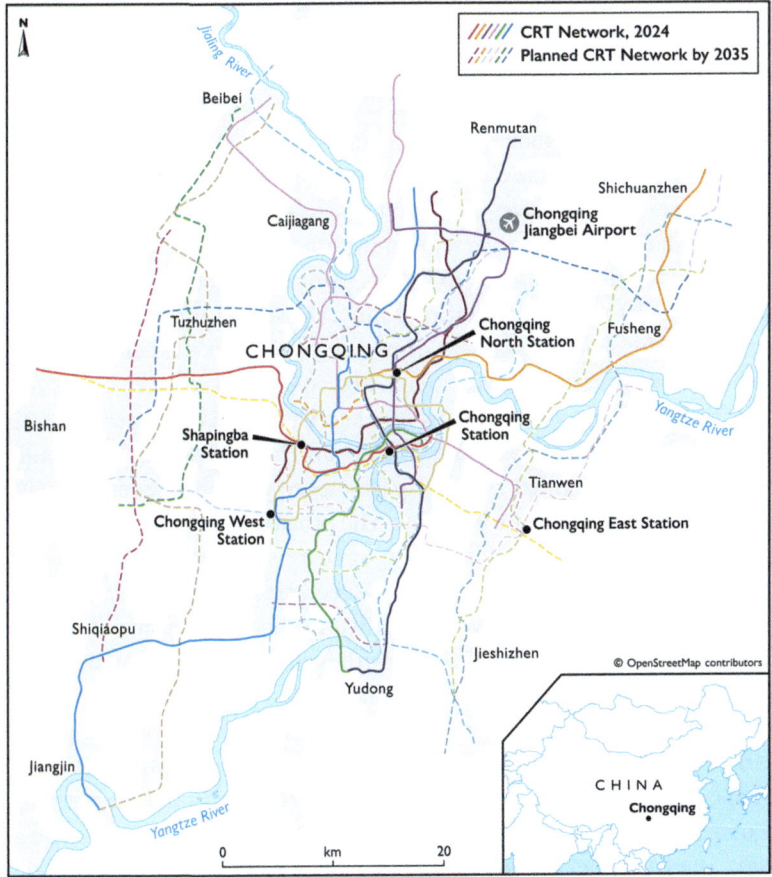

**Figure 10.1** Current and planned Chongqing Metro network by 2035.

growth. New neighbourhoods are planned with good access to the public transport system. Urbanisation has increased massively in scale since economic reforms were introduced in China, post-1978, facilitated by huge public investment in infrastructure, industrialisation and education. This 'state-led entrepreneurialism' is also associated with increased individual consumerism (Wu, 2022), hence, there is a particular form of state intervention leading to economic growth and materialism. City authorities use land sales to generate income, leading to capital accumulation, at the societal and individual levels. But, also, there is a wider social remit in terms of seeking to encourage higher-quality living conditions and improving individual well-being. Hence, the focus for city development is not solely on profit maximisation (Wu et al., 2022). City and national taxes are used to fund infrastructure, with local business taxes (40 per cent of businesses taxes remain for local use in the city) combined with national funds. A 10-year infrastructure strategy, periodically updated, is used to coordinate the projects. Wealthy cities such as Chongqing can fund many expensive infrastructure projects, hence network development is speedy and extensive. Even though the topography of Chongqing makes construction expensive, for example, some of the Metro stations are over 100 metres underground, the planned investments are extensive. Upcoming projects include extensions to the existing Metro network, with seven new projects in construction, including a new western extension to Line 4, extensions to Line 5 and Line 9, a new Line 10 and the orbital ring line. The aim is for Metro and rail to account for over 50 per cent of passenger trips in the central urban area. A new tunnel is planned to link high-speed rail services across the city, west-east. In addition, nearly 300 kilometres of cycle lanes and 600 urban parks have been constructed, ensuring that 40 per cent of the built environment is provided as open space, hence there is a renewed focus on active travel and leisure activities. The provision of active travel facilities, however, can be much improved, such as with cycle and pedestrian networks along the riverfronts, and indeed across the city. The shaping of the public transport network is rooted in the mechanisms of power, including the close political linkages with Beijing and the urban development and taxation systems that allow significant infrastructure projects to be planned and implemented. A recent national government publication in 2024 has halted the era of massive rail transit construction in China, prohibiting borrowing for infrastructure as a response to national economy and debt concerns (Liu and Zhou, 2024). But, over recent decades, the travel behaviours of the public have been shaped, in governmentality terms, to be more reliant on public transport. The exercise of power, in this case of transport planning, has been positive, shaping high levels of patronage for public transport in a rapidly growing city, hence aligning with environmental and social public policy goals.

**Figure 10.2** Extensive Metro network. The APPARATUS of the governmental institutions, techniques and administrative procedures are used to develop the Metro network across the metropolitan area – giving a productive demonstration of power by the city authorities. The Metro extends to 10 lines, linking the historic city centre to suburban neighbourhoods, including the high-speed railway stations and airport. Travel behaviours are influenced, with high volumes of passengers using public transport for everyday travel.

**Figure 10.3** Integrated urban planning and transport infrastructure at Liziba station. Metro Line 2 runs along the mountainside and overlooks the River Yangtze. Liziba station is built within an apartment block and the Metro runs straight through the building.

**Figure 10.4** Chongqing HSR, part of the national network. Chongqing is connected to other cities across the region and nationally by HSR – with more than 40,000 kilometres built since 2008. The approach to GOVERMENTALITY (administering of the population) means that cities across China are connected by HSR and social integration and cohesion are encouraged.

**Figure 10.5** An extensive HSR station in Chongqing. There are huge spaces for arrivals, waiting and departures, allowing efficient movement of large volumes of passengers.

## Connecting to the national high-speed rail network

Similarly impressive, with a different spatial extent and scale of planning and implementation, is the development of the national system of high-speed rail (HSR). The HSR network extends to over 40,000 kilometres across China, all of it built since 2008 (Figure 10.6). This is a far greater length than the HSR network operating across the rest of the world. The first HSR line in China opened between Beijing and Tianjin, coinciding with the Beijing Olympics, and the national network length is expected to reach 70,000 kilometres by 2025. Trains operate between 200 and 350 kilometres per hour (kph); consequently, travel times by train between cities have been significantly reduced and passenger capacity has also massively increased. There were nearly two billion passengers using the HSR system per annum by 2017 (Lawrence et al., 2019). Ticket prices are subsidised at low levels, competitive with air and bus, and enable most income groups to access the services. Services are frequent, comfortable and punctual, facilitating travel for business, visits to family and friends and wider tourism. Since 2013, the network has been planned, managed and operated by the National Railway Administration (NRA) and the China Railway Corporation (CRC). Both are state-owned organisations with links to design institutes and engineering and manufacturing groups, which are also partly state-owned. The Chinese HSR network has

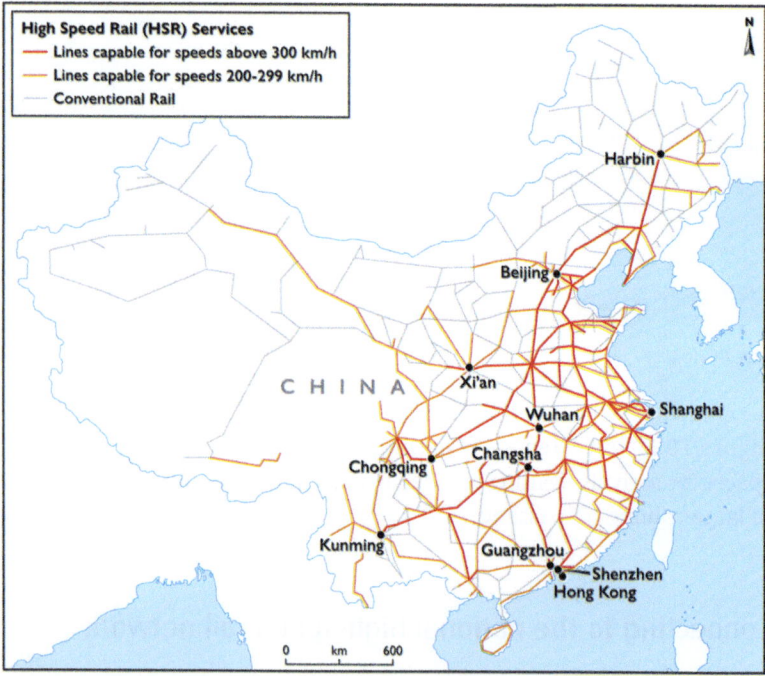

**Figure 10.6** HSR and conventional railways in China. Source: based on https://en.wikipedia.org/wiki/List_of_high-speed_railway_lines_in_China. Creative Commons.

been built at a cost of $17–21 million per kilometre, two-thirds the cost of other national HSR networks, despite many engineering challenges in crossing difficult topographies and even climates (Lawrence et al., 2019). The rail technology was initially imported from manufacturers, such as Bombardier, Alstom, Siemens and Hitachi, but is now produced by internal Chinese manufacturers. There is significant export of engineering expertise to wider systems abroad. Hence, the development of the HSR network has led to leading-edge national expertise, associated with highly skilled employment.

The HSR network is planned through the Medium and Long-Term Railway Plan (MLTRP) which is, again, periodically updated and leads to consistent project planning and delivery. There is a large technocratic resource, including technical capacity at the implementation agency, coordination with national and local spatial planning strategies, and commitment from national and local governments. The city authorities often donate land as part of their contribution to rail network development. Centralised funding, planning and approval allows speedy planning processes, much quicker than in other countries. However, this, of course, overlooks local community and wider actor participation in project planning. The HSR network has been developed for varied political reasons, including to facilitate urban development and economic growth, to allow movement within and between multiple cities and city-regions, and to integrate diverse cultures within the country. Hence, the political objectives for HSR are wider than simply facilitating increased mobility. The HSR technology is ideally suited to China, with the number of large cities (>500,000 population) and extensive journey distances between cities (200–500 kilometres and more). There are also important dimensions for national politics in demonstrating economic power, technological expertise and modernisation within China and internationally.

Within Chongqing, there are three HSR rail stations, which are used to connect to multiple cities across the HSR network, such as Chengdu, Kunming, Guangzhou, Shenzhen, Beijing and Shanghai. The HSR stations are vast, similar to modern airport terminals, with separate waiting, departure and arrivals halls, and access to multiple modes (Metro, bus, car, walk) for ongoing journeys (Chen et al., 2014). Huge volumes of travellers can pass through the rail stations, all with pre-booked tickets and reserved seats, and usually with a simple scan of an ID card to travel. In Chongqing, the HSR stations are mostly located on the edge of the urban area, similar to other cities, mainly to avoid the demolishment of existing communities and reduce building costs, but also to ensure quicker journeys between cities. New neighbourhoods are developed around the stations with good connections by Metro into the existing city centre (Chen and Hickman, 2020). Hence, eventually, the station becomes part of the city as the metropolitan area is expanded. The recent history of transport planning in Chongqing is to use public transport to facilitate urban growth, with the Metro as its central element. Urban development is used to fund the infrastructure development. Fares are subsidised, allowing relatively equitable access to the transport systems. This reflects a process of governmentality, including a technocratic-led and careful integration between transport and city planning.

# 11
# London, King's Cross

## Redeveloping the station area

King's Cross, located in the northern part of inner London, UK, has been redeveloped as a new central area neighbourhood, including refurbishment of the railway station and former industrial buildings in the surrounding area. This is seen as an example of progressive transit-orientated development (TOD), including a relatively high-density built form for the London context. The redevelopment incorporates offices, retail, restaurants, bars and apartments in the warehouse buildings, mixed with modern new buildings and public spaces, surrounding the high-quality railway station extensions and refurbishments at King's Cross and St Pancras stations (Bishop and Williams, 2016). The high-quality buildings and public spaces, including some well-known office occupiers, make this an attractive contemporary neighbourhood in London. Hence, the development is frequently viewed as a productive use of institutional power, facilitated by city authorities and implemented by developers. But, beyond this, there are concerns over the type of development that has emerged, with expensive residential and commercial units, exclusive to only higher-income residents, employers and visitors (Edwards, 2010). Hence, there is debate over the exercise of power and type of development produced by the key actors in the redevelopment process. The resulting built form is replicated in many of the larger urban development projects across London, indeed in many international contexts.

## The shaping of redevelopment at King's Cross

Historically, King's Cross was a low-lying, marshy edge of inner London, where the very poor and (as denoted) 'semi-criminal' classes lived. It was also the location for a smallpox and fever hospital. Euston Road opened in 1756, providing a northern highway bypass for London, and Regent's Canal was built in the 1820s. King's Cross station followed in 1852, providing the London terminus for the Great Northern Railway; and St Pancras opened in 1866, as the terminus for the Midland Railway. King's Cross station and the adjacent Great Northern Hotel (1854) were designed by Lewis Cubitt and the more imposing Midland Grand Hotel (1876) was designed by George Gilbert Scott. The two stations were built separately as they were owned by private railway companies, though now they are part of the same national railway network, following nationalisation of the four large railway companies in the UK in 1948. The rail network has survived as publicly owned under Network Rail, although the rail operators were re-privatised in 1997. From the 1850s, the area behind King's Cross station was used as a goods and freight yard and there were important buildings for storage, such as the Granary Building and East and West Coal Drops. By the 1980s, much of the area was derelict, following the containerisation of rail freight, and was known for crime and prostitution, but also used by some well-known nightclubs such as Bagley's and The Cross (Bishop and Williams, 2016). The area was difficult to redevelop as it was constrained by the railway lands, had much derelict infrastructure and was generally unattractive for developers and businesses. There is a fairly dense population in adjacent neighbourhoods, including lower-income residents and public housing provision, such as found in parts of Somers Town and Barnsbury.

Following many abortive attempts at redevelopment, planning negotiations for the King's Cross masterplan started in 2000, with outline planning permission given in 2006. The King's Cross station area redevelopment project was one of the most complex urban renewal projects in London, indeed in the UK, covering 27 hectares to the north of the station. In 2008, Argent, London & Continental Railways and DHL formed a joint partnership to act as a single landowner, known as King's Cross Central Limited Partnership (KCCP). Argent acted as the property developer and Allies and Morrison were commissioned as lead masterplanners. The project had significant financial backing, with Hermes Investment Management (the British Telecom pension fund) and AustralianSuper (an Australian pension fund) supporting Argent. King's Cross station was extended and refurbished, with an expanded Western Concourse providing improved passenger waiting facilities, opening in 2012 (Bishop and Williams, 2016). The neighbouring St Pancras station was refurbished in 2007, acting as the terminus for High Speed One and Eurostar trains to Paris, Brussels and mainland Europe (Figure 11.1).

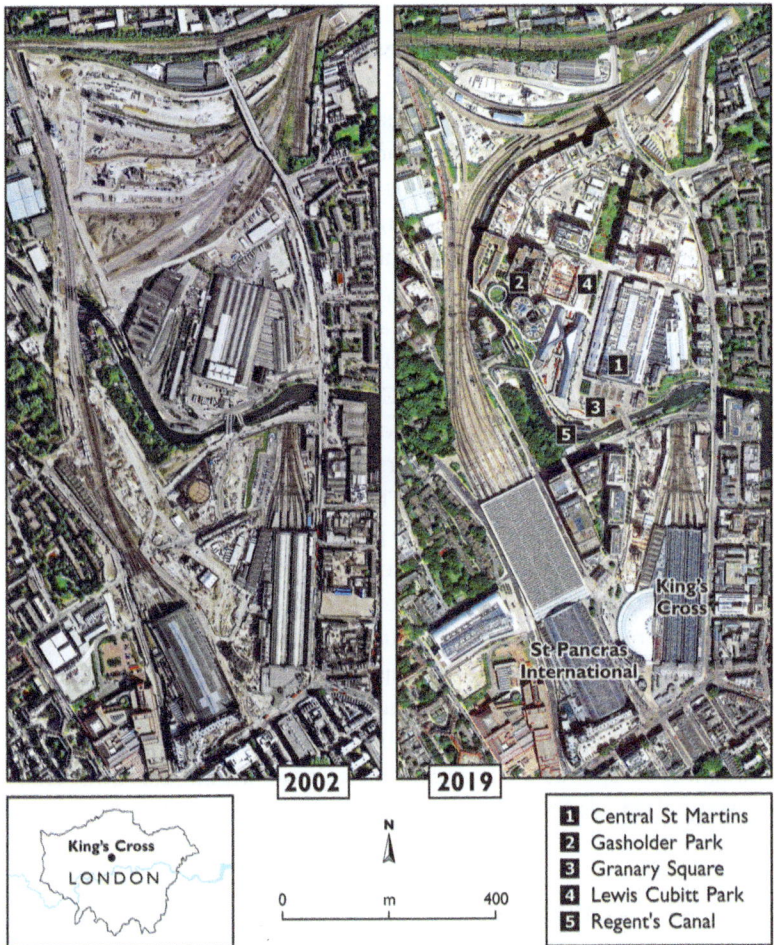

**Figure 11.1** King's Cross station area redevelopment. Source: aerial photography from Google Earth, 2020.

Redevelopment of the station area commenced in 2007 and is continuing almost two decades later. In 2024, for example, new offices for Google were nearing completion immediately to the north of the railway station. The area has been transformed into a large mixed-use development, including refurbishment of the old warehouses and industrial buildings. King's Cross is self-described as a 'benchmark for regeneration and development' and 'London's boldest neighbourhood' (King's Cross Central Limited Partnership, 2021, 1). The Granary Building has been refurbished as the home for Central St Martins, which is a renowned university for arts and design in the UK. The Francis Crick Institute and Alan Turing Institute, new university research centres, are also located nearby, hence there are well-known educational occupiers. The neighbourhood is also advertised as a new 'technology quarter', home to employers such as DeepMind, Facebook, Google and Samsung. The Guardian Newspaper's head office is adjacent to York Way. The two historic railway hotels, the Midland Grand and Great Northern, have also been refurbished into high-quality hotels (Bishop and Williams, 2016). There are around 50 new buildings, nearly 2,000 new homes, 26 acres of open space, 10 new public parks, and a range of gardens, squares and new urban spaces, such as Lewis Cubitt Park and Square, Gasholder Park and Granary Square (King's Cross Central Limited Partnership, 2021). Varied architects were used to produce a mix of architectural styles across the neighbourhood, such as Heatherwick Studio at Coal Drops Yard and Wilkinson Eyre at the Gasholders. The buildings are certified as carbon neutral and the energy supply is 100 per cent renewable, hence there is a good environmental performance from the built form. Car usage is kept at very low levels with excellent access into the public transport network and there is relatively little car parking provision, though it is difficult to find exact numbers on car parking spaces. Much of the area is pedestrianised, including Granary Square and Coal Drops Yard. The neighbourhood website suggests accordingly that this is 'the canal-side shopping and dining hotspot at the heart of King's Cross, just minutes away from the station' (King's Cross Central Limited Partnership, 2023).

## TOD and social equity?

For some commentators, the area is perceived as an example of progressive TOD development, and certainly the quality of urban development is high. But, beyond the promotional narrative, there are concerns over the form of development, including the resulting neighbourhoods and for whom they are intended and used. King's Cross is not on the scale of Canary Wharf (built from the 1980s onwards, under the government of Margaret Thatcher) or even Vauxhall-Nine Elms-Battersea (built from the 2010s onwards), and there is less controversy over the built form of King's Cross. Yet, the vast majority of its residential properties are very expensive to purchase, the offices for employment have high rents and, similarly, the retail and restaurant outlets cater for high-income visitors. The area is hence used by predominantly wealthy residents and visitors, excluding significant proportions of the population. These affordability concerns are not unique to

**Figure 11.2** King's Cross Western Concourse. The redevelopment of King's Cross and St Pancras stations is seen as progressive TOD, with very high-quality built form and public space, including redevelopment of the surrounding former industrial areas.

**Figure 11.3** Granary Square and Central St Martins. The station area redevelopment provides an attractive mix of refurbished old industrial and contemporary new buildings, with new public spaces and landscaping. The public spaces are of high quality and much of the area is pedestrianised, with little car usage. Most people travel to and from the neighbourhood by public transport and walking.

**Figure 11.4** Coal Drops Yard. King's Cross is redeveloped, but there are concerns over the form of BIOPOWER in terms of the type of development and the gentrification impacts that follow – capital growth becomes the main driver for redevelopment.

**Figure 11.5** Residential development at the Gasholders. The 'connaissance' (the corpus of KNOWLEDGE within urban and transport planning) concerning the design and implementation of major redevelopment projects in London, is understood and framed in a specific manner. Planning processes produce 'Opportunity Areas' and exclusive new neighbourhoods, aimed at higher incomes and capital accumulation, supported by excellent access to the transport network. Transport planning overlooks these concerns of affordability, deeming that all development is 'positive'.

London, indeed are evident in many other cities where major redevelopment projects have been produced with exclusive consumers in mind. In London, aspirations for affordable housing, for sale or rent, have failed to be realised and expensive housing and corporate office space dominates. The King's Cross redevelopment area was planned to deliver 750 affordable homes (40 per cent of new units), including 500 available for social rent. This was subsequently 'renegotiated' to deliver 637 affordable homes. The process of assessing 'viability' for the developer allows affordable housing targets to be reduced. More critically, 'affordable housing' is defined by the Greater London Authority as representing 80 per cent of market rent. This, of course, is not affordable to average or lower incomes, so remains fairly meaningless as a working definition (Elliott and Earwaker, 2021); even the affordable housing produced is unaffordable to many people.

This form of development, including gentrification and displacement, with subsequent social segregation, has become a redevelopment

**Figure 11.6** Office development, landscaping and the view into the city centre. The 'savoir' (the KNOWLEDGE in general) is to accept this type of redevelopment, despite significant concerns on affordability for living and working in London.

strategy internationally. Weber (2002) sees this as the reshaping of urban development processes with a primary focus on capital investment and the extraction of profit. Transport is implicated in the process, including investment in public transport, as part of the enhancement of cities, including as spaces of consumption. This can be interpreted as city planning contributing to the neoliberal effort (Culver, 2017). Smith (2002, 427) adds that there are emerging forms of governmental apparatus involved, with the state often acting in partnership with the private sector, becoming a 'consummate agent, rather than regulator, of the market'. The planning discipline in the UK has been weakened over decades, with limited strategic planning mechanisms and limited resources for public sector planning. Planning is sometimes demonised as bureaucratic and constraining housing and other urban development implementation, including by politicians and the public. But, in reality, it is the lack of public sector housing and land banking by the private developers that leads to limited urban development, or urban development of a particular form. Hence, there are important dimensions of power and exclusion in play, often supported by weak and compliant governmental actors. Swyngedouw et al. (2002) warns that neoliberalism, as part of the process of globalisation, shapes new urban developments in order to maximise capital, meaning that more socially orientated public policy goals are overlooked. Sometimes, governance arrangements are even rearranged to aid implementation processes, such as using 'special delivery' arrangements, which also weaken participatory mechanisms. Consider the so-called 'Opportunity Areas' in London, and their production of new, luxury developments, unaffordable to many. Over decades, the urban planning system has become complicit in this type of inequitable development. Contemporary planning processes are led by the developers, assisted by selected urban planning, urban design and transport-planning consultancies. The Greater London Authority and Transport for London provide 'support' for new development projects, but only as 'captured' actors and partners in the redevelopment process. The apparatus of the major governmental institutions are compliant, promoting the redevelopment plans through the spatial plan (Greater London Authority, 2004) and transport strategy (Transport for London, 2018). Transport for London and transport consultants provide analytical reports to support the envisaged growth of the area. The local borough councils (the London Borough of Camden in the case of King's Cross) and local communities are unable to gain sufficient support for their views on the shape of redevelopment. There is some controversy and debate, but, ultimately, the redeveloped neighbourhoods largely reflect the views of the developers – and this state of planning practice remains unquestioned. This is described as a 'new choreography of elite power' (Swyngedouw et al., 2002), resulting in increases in social polarisation at the city level and is purposively overlooked by the governmental actors.

Hence, the forms of redevelopment and gentrification experienced in London have evolved from those seen in the 1960s (Glass, 1964) and even the urban development envisaged in the urban renaissance movement, the state-sponsored urban renewal programme in the UK, adopted from the early 2000s onwards (Urban Task Force, 1999). This was the then Labour government of Tony Blair attempting to reverse urban decline through the renewal of cities, with increased urban densities and mixed uses encouraged in redeveloped urban areas, including a focus on transit-orientated development. The objectives were to improve social cohesion in urban areas, but also avoid suburban sprawl and development on greenfield sites. The urban policy led to the classification of 'Opportunity Areas' and implementation of major redevelopment plans, such as seen at King's Cross. There was also the dimension of promoting London as a 'World City', predicated on economic and urban growth. Over time, this emphasis on coordinated, large-scale urban redevelopment has increased, with London becoming an outlet for international residential property consumption and global capital flows (Smith, 1987). The well-meaning public sector, including urban and transport planners, have struggled to defend local community interests and wider public policy goals. Developers ensure that the profit motive maintains the shape of new development. Gentrification, displacement and exclusion have become much more systematic and overwhelming (Hackworth and Smith, 2001). Many urban and transport planners are employed in the extensive market of private consultancies, which directly support the private developer interests. Hence, transport planning has become embroiled in the process of maintaining capital growth through urban development, yet ignores the political implications of its actions. Harvey (2005, 2) defines neoliberalism, in this context, as

> a theory of economic practices that proposes that human well-being can best be advanced by liberating individual entrepreneurial freedoms and skills within an institutional framework characterised by strong private property rights, free markets and free trade .... Furthermore, if markets do not exist (in areas such as land, water, education, health care, social security, or environmental pollution) then they must be created, by state action if necessary.

This more critical understanding of the urban redevelopment processes, as evident in London, exposes the agendas and actors at play. The complex network of actors and processes involved is not always transparent, but shapes how redevelopments, such as King's Cross, are planned and implemented. The general knowledge (savoir) and understanding on urban planning and regeneration is framed to serve the interests of the private developer, in other words the most powerful actors in the development process. The King's Cross redevelopment area may look impressive at first glance, providing a high-quality transport interchange and refurbished industrial warehouses, alongside contemporary architecture and high-quality public spaces. But, a closer examination raises many questions over the unproductive application of power. 'Regeneration' is not seen, as we might imagine, as a process to assist lower-income and existing communities, but rather as a means to facilitate capital growth, benefiting only a small cohort of key actors and the population (Edwards, 2010). Transport infrastructure provision is provided, but the relationships between the existing communities, incomers, the national and city authorities and developers reflect many tensions, and powerful interests dominate. Transport planning exacerbates inequitable outcomes, facilitating high-end development. Increasingly, and perversely, the infrastructure becomes part-funded by the development. This is a process of using biopower to regulate the production of new neighbourhoods. Included is the shaping of knowledge, in relation to urban and transport-planning procedures. There is little debate concerning the nature of the transit-orientated development areas, including the role of transport infrastructure, provided at public expense, and the land value gains that fall to key actors. There is little critical discussion of the priority given to specific urban development and transport projects over others, including the neighbourhoods that are served or not served. The major adverse impacts on social inequity remain largely unresolved, whereby rises in property prices are captured by the developers and landowners, while lower-income people are displaced by wealthier people as areas are 'regenerated' (Edwards, 2010). Any debates by the local boroughs and communities are marginalised and the voices of the displaced or excluded populations are lost. Transport planning, as a discipline, ignores all these complexities, assuming that all development surrounding new infrastructure is 'positive'. But, this is a huge simplification, leading to problematic social outcomes. Project appraisal procedures are continuously refined, yet ignore distributional issues, as if they are unimportant and transport infrastructure does not contribute to them (Hickman and Dean, 2018). The urban development and transport-planning processes, therefore, facilitate high-quality new neighbourhoods and transport infrastructure. But, unfortunately, their social impacts are often regressive and overlooked.

# 12
# Rio de Janeiro

## Beyond the congested city

Rio de Janeiro has a spectacular natural setting, but is also a city of huge contrasts and social inequity, with high-income neighbourhoods and unplanned, informal neighbourhoods (favelas) often located immediately adjacent to each other. Rio is the second largest city in Brazil, founded by the Portuguese in 1565, and with a population of nearly 6.5 million. Parts of the urban area are listed as World Heritage sites and the natural topography, beaches and landmarks make the beachfront city unique. Beyond the landscape, inequity is perhaps the most prominent feature, with high levels of income inequity, and associated crime and perceptions of crime. There is very significant differential access to transport systems and access to activities.

The city has experienced high levels of traffic congestion for decades and highways dominate many of the urban neighbourhoods. Buses and the Metro are the main forms of public transport, and there have been recent extensions to the Metro and bus rapid transit (BRT). There are also more conventional bus, trains, tramway and cable car systems. The Metro and tramway are focused on the central areas of the city and the outer areas have poorer access to public transport. The BRT system was developed as part of infrastructure investments prior to the Olympic Games in 2016 and extends to 168 kilometres in length with over 3.5 million passengers per day (Bus Rapid Transit Centre of Excellence, 2023). The public transport network, however, is still limited in coverage and not sufficient to serve the large urban area, particularly in the outer areas where many of the middle- and lower-income residents live. Journeys are often lengthy, uncomfortable and crowded, with many commuters travelling over two hours every day by public transport. The indignity of the daily commute on the crowded BRT has gained social media attention, with the buses described as 'the new slave ship' (McCann, 2022). Fare levels are perceived as high, relative to the quality of service provision, and fare rises often lead to controversy. The state of public transport is central to many public protests, alongside wider concerns over crime, social welfare, education, health and corruption (McEwen and Pimental Walker, 2015). The planning and development of the BRT lines were controversial, with the construction companies and bus operators making large profits from land development and the public–private implementation and operating model, whilst service quality remained poor. There were also problems with residential displacement and resettlement of the urban poor, including on the Transoeste, TransCarioca and TransBrasil BRT routes, as land was cleared to make space for the routes. Some of the low-income residents were displaced to the western outer areas of Rio, with the resettlement neighbourhoods often under control of militias (McCann, 2022). Differential access to public transport remains across the city, making participation in employment and other activities difficult for lower-income and other disadvantaged groups. The building of new public transport systems is hence poorly implemented and actually perpetuates social equity problems in some cases. The policy circulation and transfer model, with BRT inspired by practice in Curitiba and Bogotá, seems to break down as contextual differences mean that new infrastructure does not work as envisaged, and policy models mutate rather than simply transfer (Peck and Theodore, 2010). Transport planning is, of course, embedded in social structures, and where social inequity is so stark, it is easy for power to be applied with unintended and adverse impacts.

## Praça Mauá and Rio Branco streetscape improvements

Some further innovative projects have been developed to help improve the transport system in the central areas of the city, including the removal of the urban highway at Praça Mauá and investment in public transport and public space improvements along Rio Branco, part of the wider Porto Maravilha (the beautiful port) redevelopment project (Figure 12.1). But, similarly, they have suffered from problems in implementation and social impact.

Praça Mauá is a public square found in Rio de Janeiro's historic Centro commercial district, linking Rio Branco Avenue and Porto Maravilha. The historic port area originates from the early 1900s and was used to bring freight ships and passengers, including migrants and

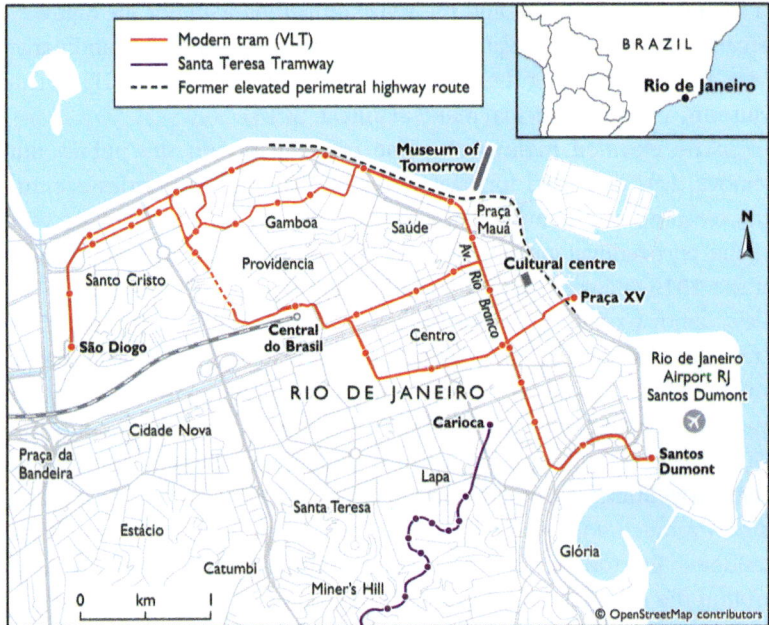

**Figure 12.1** Transport and streetscape interventions at Praça Mauá and Rio Branco, Rio de Janeiro.

tourists, to the city. It was one of the most important and vibrant parts of Rio, overlooking Guanabara Bay, with many historic buildings. These include A Noite (The Night), a high-rise building influenced by the Art Deco style and modernist movement. This was the first skyscraper built in Latin America, completed in 1922, using the new building technique of reinforced concrete. It was originally used as headquarters for the city newspaper 'The Night' and also hosted the Radio Nacional.

From 1957 to 1978, an elevated urban highway, the Elevado do Perimetral, was gradually constructed over the Avenida Rodrigues Alve, including crossing directly over Praça Mauá. The highway was built to give greater capacity to traffic movements, connecting the Rio-Niterói Bridge and the north of the city to Santos Dumont airport, the city centre, and the waterfront neighbourhoods to the south. The highway carried forty thousand vehicles per day and was seen as an important highway connection. However, it cut through many urban neighbourhoods, such as Caju, Santo Cristo and Gamboa, and created a huge noise, air pollution and severance problem, blocking views and using space on the waterfront. The surrounding areas suffered blight from the impacts of the elevated highway, and the space immediately under the highway became dangerous to visit. Important historical areas and landmarks were ruined, including the small streets around the National Historical Museum and old municipal market (Villela, 2014).

The elevated highway became unpopular with the public and removal was discussed for over 20 years. The city authorities eventually removed the highway, with demolition in 2013 and 2014, as part of the port redevelopment and wider revitalisation of Rio, in advance of the 2016 Olympic Games. A road tunnel replaced the elevated highway, removing the severance problem, but not the highway capacity. Praça Mauá was redesigned as part of the urban renewal, including a much-improved pedestrian environment and public space, with a tram line running along Avenida Rodrigues Alve and through the square. The Museu de Arte do Rio was refurbished and extended and the Museum of Tomorrow built, designed by architect Santiago Calatrava. Praça Mauá is now a much-improved public space and an important destination for residents and tourists in the city. The tram line helps with movement around the neighbourhood, with a high-quality pedestrian environment and access to street art, landmarks and historic buildings.

Rio Branco Avenue leads from Praça Mauá and has also had some streetscape enhancements. This is a historically important street in the Centro area with many architecturally significant buildings, such as the Municipal Theatre and National Library. From the 1950s onwards, the street became increasingly congested with vehicles, as car ownership and usage grew in Rio. The old historical buildings became polluted and the street was very noisy and unfriendly for pedestrians. Again, as part of the redevelopment planned around the Olympic Games in 2016, Rio Branco was converted back into a more pedestrian-friendly space. Private cars were removed and the street converted into a tram, pedestrian and cycle space in 2015. The tramway (Veiculo Leve sobre Trilhos, VLT Carioca) opened in 2016, connecting the central business district to the Santos Dumont Airport and Novo Rio terminal bus station.

## Regeneration and social impact?

Again, there is a complex picture of impact from the port redevelopment. The new streetscapes in Rio Branco, Praça Mauá and surrounds are part of the planned wider revitalisation of Porto Maravilha and provide much-enhanced public spaces. This is a huge regeneration project, being implemented over decades, but is surrounded by much debate

**Figure 12.2** The BRT system in Rio de Janeiro. Sometimes well-intentioned transport projects are poorly planned. In this case, implementation of the BRT has experienced difficulties with fare levels and social displacement along the corridor.

**Figure 12.3** Praça Mauá renewal. The Porto Maravilha redevelopment has much improved the spaces and streets along the waterfront, particularly with the removal of the Elevado do Perimetral. Praça Mauá has become a popular place to visit again, with an attractive pedestrian environment. Source: Matt Aucott (UCL).

**Figure 12.4** The Porto Maravilha redevelopment project has created new commercial and residential developments, but has also been controversial relative to who it serves. There is an issue of GOVERNMENTALITY in that much of the new development is serving higher-income residents, businesses and visitors.

**Figure 12.5** New tramway in Porto Maravilha – this has given improved public transport access to the area and the streetscape is well designed and attractive. Source: Matt Aucott (UCL).

**Figure 12.6** The favela of Morro da Providência gains little from the surrounding development. Porto Maravilha has been redeveloped, but the lower-income groups have largely been excluded, as the new activities and transport infrastructure do not serve their needs. This represents the governance of people by others (BIOPOWER), extending beyond the administration of the people through urban planning, including how social segregation is furthered. Source: Matt Aucott (UCL).

and controversy. The plans began in 2009 as part of the urban renewal element of the World Cup in 2014 and Olympic Games in 2016. There has been associated corruption and misspending of public funds for the public events, including the priority given to funding and hosting sporting events in view of widespread urban poverty. The waterfront urban renewal attempted to use the public events to redevelop decaying urban neighbourhoods, following the example of Barcelona and the transformation of the dilapidated Port Vell alongside the 1992 Olympic Games. The plans for Porto Maravilha are impressive in scope, aiming to reconnect the historic waterfront to the surrounding neighbourhoods, with new commercial development, homes, offices and leisure space. The area is being repopulated again, with the port area district planned to accommodate upwards of 100,000 people relative to its current 32,000 population (Keeling, 2015). But, there are huge complexities with a project of this size, and in a city such as Rio. Projects are difficult to implement and to complete on time and to cost. More importantly, there are social equity concerns, as the city is gradually rebranded for the higher-income groups and visitors, yet the low-income groups lose out as their everyday experiences are not improved. For example, Morro da Providência, Rio's oldest unplanned favela, sits adjacent to the port area. It is likely that the low-income residents will be displaced to outer parts of the city as land values rise. The favelas are surrounded by new infrastructure, but gain little from it (McEwen and Pimental Walker, 2015). Again, the urban planning and regeneration efforts overlook significant cohorts of the population. In terms of transport, funding is scarce and there are concerns that central urban projects gain most focus, at the expense of investment in transport modes that will be used by lower-income and disadvantaged residents on the edge of the metropolitan area. The major sporting events are used to give funding and provide deadlines for redevelopment, but often the priorities for investment, including the focus on redevelopment that leads to increased land values, gentrification, displacement, real estate speculation and financialisation of development processes, are controversial. The removal of the Perimetral and building of public transport projects were progressive, reducing the priority given to the private car, at least at street level. But, the distributional impacts of transport interventions and the associated urban redevelopment need much more serious consideration if social equity problems are to be resolved, or at least reduced. These types of major area redevelopment and infrastructure projects are hugely complex to plan and implement. There are strong dimensions of power in terms of the resulting impacts and the new neighbourhoods are often not used by the incumbent, lower-income populations. On purpose, or by default, the projects are targeted at higher-income groups, serving the areas that they work in or visit, and facilitating the activities and lifestyles that they aspire to. The external image of the city may be improved, yet the process of biopower does little for the very apparent problems of poverty and deprivation. The apparatus maintains the current social status – and much greater reflection is required to consider how urban and transport planning can change the social inequity in income and activity participation, in a more positive way.

# Part V
# Discontinuity

## Towards more sustainable transport systems

The principle of discontinuity is used to examine the historical trajectory of social practices, and in particular to undermine the status quo (the existing state of affairs) (O'Farrell, 2005). For example, Foucault rejects continuity for varying social practices, arguing that the conventional practices are unjust (Foucault, 1961, 1975, 1978a). In transport planning, changes in the forms of order are evident over time, including towards more environmentally and socially equitable practices and states of being. There are procedures involving the examination of historical trends and estimations of causes and effects, used to explain current behaviours and change. Historical trends are often extrapolated to estimate future trends. For example, forecasting is used as the primary procedure to estimate transport patterns, what is likely to occur and how these should be responded to. Certain objects, categories, classifications and relationships are used to describe the current state, with relationships often viewed as static over time. There are also assumed, but largely undiscussed, 'end goals' for transport policy, such as the requirement for increased mobility, often linked to economic growth. Certain rates of motorisation seem to be aspired to, for example, that motorisation and mobility levels experienced in the USA can be 'achieved' in all societies. Rarely are there significant critical and transparent debates on concepts of growth, such as economic, urban and mobility growth, or how these might relate to environmental or social goals. A Foucauldian perspective, therefore, can be useful here, to more critically examine the public policy positions taken, with human behaviour, experience and practice understood as an event or series of events, within the wider structure of society. The aspiration is to understand how discontinuity has been and can be achieved, that is, how travel behaviour can be reinterpreted and reshaped.

Problematisation is a key concept to help facilitate change, allowing an understanding of current practice and its associated problems, to make these issues more transparent; and, hence, lead to different understandings of social practice. Foucault does not only write the history of a period, but also considers how practices were accepted and might be open to change. He attempts 'to incite new reactions and induce a crisis in the previous silent behaviour' (Foucault, 1983b, 74). In transport, this is most likely in terms of problematising the dominance of motorisation; the lack of funding and prioritisation for other modes of travel; the poor integration of transport with urban planning; the low resource given to urban planning; and, as an output, the poor progress against environmental and social goals. The task is to defamiliarise the denoted 'common sense' (Crotty, 1998), for example, the perceived 'practical' and 'sound' judgement on transport systems, including that transport systems should be dominated by the car and streetspace should be mostly given to the car to encourage mobility and economic growth or accommodate complex modern lifestyles. This allows new approaches and actions to emerge, including much greater investment in public transport, walking and cycling, alongside compact urban planning. Examining problematic social practices and conflicts with policy goals hence becomes important so that new forms of practices can emerge.

Current transport systems can often be challenged, and often continuity is simply a justification of current practice and its associated adverse impacts and injustices, or maintenance of the status quo for the current 'winning' actors. Foucault developed his critical historical perspective on social practices to highlight their absurdity. He often focused on prescriptive texts, creating the impression of a 'natural' or 'perfect order', including how people ought to behave and society be constructed. The position in transport is, similarly, often absurd, with the dominance of motorisation spreading across multiple and varied contexts, despite many significant adverse impacts. This would be seen as a contradiction by Foucault (1969); understanding discourse helps reveal and expose these conflicts, and the apparent cohesion in approach can be challenged. The difficulties in providing high-quality public transport, walking and cycling facilities in cities and wider regions seem widespread, yet public spending remains challenging to justify and even fairly marginal reallocations of streetspace can prove contentious. As Foucault (1969, 6) tells us 'The history of thought, of knowledge, of philosophy, of literature, seems to be seeking, and discovering, more and more discontinuities, whereas history itself appears to be abandoning the irruption of events in favour of stable structures'. There are parallels to be made with transport planning, where there are contradictions between thought (where we might like to be) and history (the current travel patterns). There is much inertia, and dominant actors profit from the current status. But, the requirements of environmental and social sustainability increasingly demand different transport strategies and projects to be implemented. These issues are examined with six case studies, each demonstrating key periods of discontinuity within their particular transport and urban planning contexts, leading to radically changed travel behaviours. These are Utrecht (transition from car-based system to cycling and public transport), Copenhagen (development of a cycling city), Malmö (focus on the space between buildings for social interaction), Dar es Salaam (bus rapid transit in a rapidly growing city), Delhi (regional Metro) and Shenzhen (electric bus fleet).

# 13
# Utrecht

## Problematising travel behaviours

For many cities, cycling is, as yet, a marginal activity, perhaps accounting for 1–5 per cent of trips. But, there are some notable exceptions, particularly in the Netherlands, where cycling is the major mode of travel on a trip basis for many cities and smaller urban areas. The Netherlands has a population of 17 million people and over 22 million bicycles. But the Netherlands purposively chose to make cycling central to its transport planning – it did not just happen. The national and city authorities have continually improved cycling facilities and cycling experiences over the last five decades (Pucher and Buehler, 2008, Oldenziel and Albert de la Bruhèze, 2011). This strategy arose in response to the energy crisis in the 1970s and the high traffic casualty rates, particularly involving children as pedestrians. There was a public debate around increasing motorisation and the associated adverse problems, hence a critical stage of problematisation concerning the impacts of motorisation and alternative, more environmentally and socially acceptable means of travel. Cycling, walking and public transport were encouraged as the priority modes of travel through the transport systems, supported by compact and well-connected urban centres.

## Improving the cycling environment

An example of a city with excellent cycling facilities is Utrecht. This is the fourth-largest city in the Netherlands, with a population of 348,000. 125,000 persons cycle every day in the city; 51 per cent of all trips and 59 per cent of trips to the city centre are by bicycle (Netherlands Institute for Infrastructure and Water, 2018). There are many different types of bicycles in use, mostly the Dutch classic bikes, but also cargo bikes, OV-fiets (public transport hire bikes) and many others of different styles and sizes.

Utrecht is progressing many innovative streetscape redesign projects, consistently improving the pathways and facilities for cycling and walking and improving public spaces over time (Figure 13.1). Much of the city centre was dominated by urban highways, originally built in the 1970s, but these have gradually been removed to improve the urban environment, including the radial routes and the inner ring highways that surround the city centre. An important example is the Croeselaan, previously a dual carriageway for traffic, which provided a through route around the city centre leading to a multi-storey car park. The route has been completely redesigned so there is no through traffic beyond the Jaarbeursplein. This now functions as the western station square, and the car park has been placed underground. The reduced traffic flow meant that the corridor could be used differently – there is just one lane of traffic in each direction, and the remaining space is used for cyclists and pedestrians. There is a high-quality, bi-directional wide cycle lane (4.5 metres wide) and a linear art park placed in the central median. The park has art, sculptures, mature trees, landscaping and places to sit and relax. The street lights are solar powered and even the street signs are made from bamboo (Bicycle Dutch, 2019b). The Jaarbeursplein is now one of the main public spaces in the city, with the steps leading up to the station providing a view of the cyclists, pedestrians and social activity. Vredenburg is the busiest cycle route in Utrecht, providing the link between the station and the historic city centre, with an average 33,000 cyclists using this single street every day.

One of the contemporary challenges in the Netherlands is providing sufficient space for cycle parking, particularly in urban centres and at central stations. Utrecht's stationsplein (station square) cycle parking facility opened in 2019, designed by Ector Hoodstad Architects, and has space for 12,500 cycles and 1,000 OV-fiets (public transport hire bikes). This is a huge cycle parking space, built over three floors, under the station square linking Utrecht Central station to the neighbouring shopping centre and historic city centre. It is open 24 hours a day, seven days a week, providing free cycle parking for the first 24 hours with an electronic checkin via the OV-chipkaart (public transport smartcard). Thereafter, it costs €1.25 per 24-hour period for standard bikes and €2.50 for irregular-sized bikes. An annual subscription is €75 for standard bikes. The lower and upper floors (levels -1 and +1) are for cyclists leaving their bikes for up to a day and the ground floor (level 0) is for people with

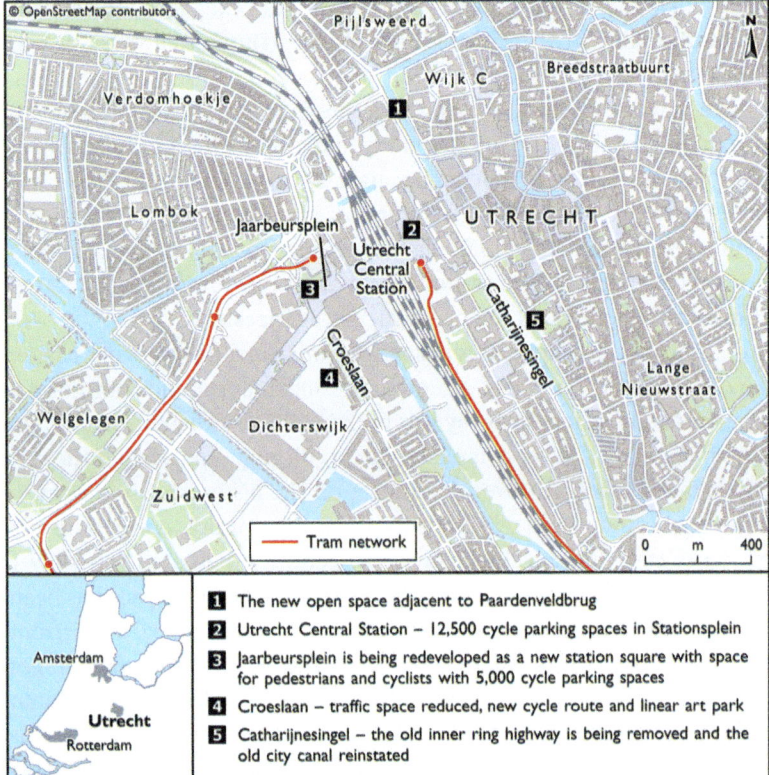

**Figure 13.1** New cycling facilities and public spaces in Utrecht.

parking subscriptions, which allow longer stays (Bicycle Dutch, 2019a). Many regular commuters leave a bike here, and use it for the 'last kilometre' part of their trip, either to/from home or work. This is often how regional, inter-city journeys are made – by rail and bike, designed for seamless integration. There are special parking areas for oversized bikes, such as cargo bikes and those with baskets or child seats; and also a cycle repair workshop.

Cyclists enter the parking facility by entrance ramp and ride past the 'p-route fiets' space indication system. The messages inform visitors where the nearest cycle parking space is, how many hire bikes are available and when the train is departing. There are 161 digital signs, either at the entrances and exits or on each parking bay, and also there is an associated App which gives information on free cycle parking spaces in the city. The parking facility is supervised by around 40 service staff who ensure bikes are parked correctly and are not left for longer than

28 days. The design and building of the facility is a collaboration between the Municipality of Utrecht, ProRail and NS (Dutch Railways). It was built over five years, and opened in stages, with a cost of €30 million, a large amount for cycle parking. But, on entrance, it is obvious this is an excellent facility, using high-quality materials including stone, wood, glass and steel. It is a welcoming, light and very well-designed cycle park, developed with the cyclists' comfort in mind. Beyond the stationsplein, there are a further 20,000 cycle parking spaces in the surrounding area, associated with the station and major offices, including 5,000 at Jaarbeursplein, 3,000 at Knoop and 1,240 at Sijpesteijn.

A further project is to reinstate the city canal by removing part of the old inner-city ring road at Catharijnesingel, north and south of the city centre. The urban highway was built in 1973 and had up to 11 lanes of traffic, but was never completed into a full ring road around the city centre due to concerns over the impact on the historic environment. It was actually underused relative to its projected capacity, as the route was not completed and development of the central urban area was slower than envisaged. The highway was closed in 2010 and the space has been reconverted back into the old city canal, providing a new waterfront park in the city centre (Bicycle Dutch, 2016). The car parking garages formerly alongside the highway have been removed or relocated into underground garages. The new open space is now used by pedestrians and cyclists and adds to the improved public spaces and cycle networks being developed in the city. At one of the renewed open spaces, adjacent to Paardenveldbrug, there is an open-air bar and barbeque where people drink, eat and socialise, overlooking the new waterfront.

New urban neighbourhoods, such as Leidsche Rijn and Merwede, have been built or are being planned, again with high-quality public transport, walking and cycling provision, and quick connections into the city centre of Utrecht. Merwede will be home to around 12,000 residents and is designed as a virtually car-free neighbourhood, with 20,000 cycle parking spaces and only 250 shared cars available for residents.

## Cycling as a social norm

Cycling has become a dominant means of travel in the Netherlands, enabled by the cycle pathways and facilities that have been provided across the country. There are 37,000 kilometres of cycle paths and the Dutch people cycle on average 880 kilometres per year with over 300 journeys per person – an incredible volume of cycling. The social practice of cycling starts in childhood; children typically sit in a child seat on their parent's bike, learn to ride when four or five years old, and cycle to school, without parents, from the age of eight to ten. Hence, cycling as an activity is encouraged and normalised from an early age and continues throughout life, including into the older age groups. The benefits of high levels of cycling and improved public spaces are very apparent to individuals, transport planners and politicians. Cycling is designed to be the quickest and most convenient mode of travel, but the activity also helps reduce obesity and the risk of some forms of cancer, cardiovascular diseases, diabetes and depression. Riding a bike every day reduces the risk of premature death by 40 per cent; it prolongs life expectancy on average by up to 14 months. And, of course, it is fun: two-thirds of Dutch residents aged over 18 associate cycling with feelings of 'joy'. They are more satisfied, less stressed, more relaxed and experience greater freedom than those who drive to work (Netherlands Institute for Transport Policy Analysis, 2018). Cycling is used for over a third of trips in the Netherlands, with a higher proportion for trips in the urban centres, and even the longer trips (7.5–15 kilometres) are now increasingly carried out by bicycle or e-bike (Netherlands Institute for Infrastructure and Water, 2018); cycling has become the ubiquitous means of travel and even an integral part of the culture of the country.

**Figure 13.2**  Central Station in Utrecht – the station and surrounding area have been redeveloped to provide high-quality rail travel and convenient interchanges between rail, bus, walk and cycle – with cycle parking provided with thousands of secure spaces. This is the PRACTICE of station redevelopment in the Netherlands and replicated in many cities.

**Figure 13.3** Rail connections are provided between urban centres in the Netherlands. Travel by rail is possible between cities, and beyond, with last-kilometre connections provided by individual or shared bicycles.

**Figure 13.4** High-quality cycle parking: parking space built over three floors and open 24 hours a day.

**Figure 13.5** Cycle path and linear park at the Croeselaan. The urban highways were PROBLEMATISED and have been reconsidered and retrofitted. A new linear art park has replaced the previous dual carriageway at Croeselaan.

**Figure 13.6** The city canal is reinstated at Catharijnesingel, north and south of the city centre, by removing part of the old inner-city ring road.

**Figure 13.7** Extensive cycle facilities. There are generous cycle pathways throughout the city and even the cycle parking has electronic signage to show space availability.

**Figure 13.8** Dafne Schippers bridge and cycle path over the school. An innovative bridge crossing over the Amsterdam-Rhine canal, extending over the top of a primary school. Cycle facilities are provided in the outer areas to allow cycling across the city.

**Figure 13.9** Leidsche Rijn new neighbourhood, built with excellent public transport, walking and cycling facilities, and very low car parking provision.

**Figure 13.10** High-quality cycling experience. Cycling becomes the NORMALISED means of travel and even an integral part of the national culture.

# 14
Copenhagen

## Cycling and liveability

Copenhagen is also a 'progressive' cycling city, perhaps with a little more self-aggrandizing than found in the Netherlands, but seeks to be amongst the world's best cities for cycling (Gössling, 2013). The city also considered urban highway building in the 1950s–70s, with new routes planned through neighbourhoods and alongside the city lakes. These were never built, mostly due to lack of funding, but the historic city tram network was removed in 1972 and cycling levels reduced to a low for Copenhagen of around 20 per cent of trips in the 1970s (Colville-Anderson, 2018).

Since the 1980s, cycling has been encouraged and the cycle network gradually extended. Mode share for cycling to work and education in Copenhagen has risen from 36 per cent of trips in 2010 to nearly 50 per cent in 2018. Twenty-five per cent of children cycled to school in 2018; there are an estimated 40,000 cargo cycle bikes in Copenhagen, and the number of cyclists entering the city centre now outnumbers vehicles. Improving the cycle network has involved a significant level of funding: two billion DKK (£230 million) has been invested in cycling facilities, including an extensive urban and regional network, the latter known as cycle superhighways, over the period 2008–18 (City of Copenhagen, 2018, Colville-Anderson, 2018). The city's history shows that cycling can be the most convenient, safe and enjoyable way to move around a city, but it has to be planned with high-quality facilities, securing funding and implementing projects consistently over years, iteratively improving the cycling experience over time. It is the range of initiatives, including the cycling infrastructure, but also regional and urban planning (for example, housing, employment and school location), traffic demand management, cycling infrastructure, parking provision, taxation and wider public policy, that together help cycling become an important means of travel and the preferred mode for many. Cycling has become a political tool for achieving a sustainable and liveable city, and the city is marketed according to its sustainability credentials: a bicycle-friendly city is perceived as a city with more space, less noise, cleaner air, healthier citizens and a stronger economy – a place where people wish to be (City of Copenhagen, 2011). This has been a clear process of discontinuity from the 1980s onwards, with a revised social practice produced around cycling.

## Speed and convenience

The aim for the cycle network in Copenhagen is to make cycling the quickest and most convenient mode of travel around the city. Climate change, wider environmental and health benefits are, of course, important public policy goals, and increased cycling is important to these, but cycling is mostly promoted through the goals of speed and convenience to the user: 48 per cent of Copenhagen cyclists say they choose the bicycle because it is the fastest and easiest way to get around the city (City of Copenhagen, 2011). An extensive network of segregated cycle pathways has been developed over the last 20 years, so that cycling has become safe and connected throughout and beyond the city. Cycle paths are generally separated from traffic by a kerb, with the cycle path higher than the traffic lane and at the same level as the footway. Cycle path width is at least 2.3 metres, and often much wider, up to 4–4.5 metres, to allow cycles to ride side-by-side, to overtake and to accommodate greater cycle volumes.

Important infrastructure projects include segregated cycle routes on streets, bridges across water, large roads, railways and routes across road junctions. Contraflow cycling is encouraged on one-way streets, together with cycling across pedestrian squares and traffic calming around schools and in residential neighbourhoods (Colville-Anderson, 2018). Many of the existing cycle routes are well used in Copenhagen and busy during peak periods, hence the cycle tracks are widened and alternative routes built to make connections between neighbourhoods easier. Cycle parking is provided in new developments, office spaces and at Metro, suburban and regional railway stations. This, as yet, is not comparable to the great cycle parking garages being built in the Netherlands. Bike rental is available across the city, such as the Donkey Republic bike sharing system. Technological innovations can also be useful where these support cycling, including 'green waves' of traffic lights for cyclists, digital bike counters, tool racks and air pumps, and even cyclist-orientated rubbish bins and junction footrests. Typical problems with low-quality cycle infrastructure are avoided, such as cycle lanes simply painted on the highway, discontinuous cycle routes, and cycle paths built adjacent to parked cars and in the way of opening car doors. Many cities start with this type of infrastructure and it has to be upgraded over time as difficulties become evident for cyclists.

Examples of high-quality cycling infrastructure include Nørrebrogade (North Bridge Street), which is part of the C95 cycle superhighway and the busiest cycle street in Copenhagen, linking the suburbs to the city centre. The street used to have two car lanes in each direction, but this was modified in 2009 to accommodate just one car lane and a wide, segregated cycle lane in each direction, this being up to 4 metres in width. Two sections of the street have a 'traffic filter', which means that cyclists, pedestrians and buses are allowed to pass, but no other motorised vehicles. 48,000 cyclists used the street at the highest daily recording in 2016, and car traffic has reduced by 50 per cent (Buczynski, 2018). In addition, Lille Langebro cycle bridge crosses the harbour at the BLOX, which is home to the Danish Architecture Centre. The bridge gives a direct connection from the city centre to Christianshavn. Further to the southwest, the Cykelslangen bridge and Brygge Bridge provide a link to Islands Brygge. The former, the Bicycle Snake bridge, allows a winding ride from the shopping mall down to the harbour, allowing a view over the city and waterfront. Together, the cycle paths, boardwalks and new bridges animate the waterfront, bringing life to a part of Copenhagen harbour.

## Wider urban planning

Urban planning is critical to cycling, as the built form can be shaped in a way that supports cycling, alongside walking and public transport. Copenhagen has a long history of using urban planning, stretching back to the 1947 Finger Plan (Egnsplan) (Knowles, 2012), which sought to guide development in corridors along the planned suburban railway lines. Each station was the focal point for high-density housing and other uses such as retailing or employment. In between the development corridors, the open space was maintained, providing 'green wedges' of space for leisure or farmland uses. This development vision was maintained through regional plans in the 1970s and 1980s, and extended with a new corridor of development to Ørestad in the 1990s (Figure 14.1). The built form can still be seen today, extending along the rail corridors and punctuated by open space.

**Figure 14.1** New urban development around the public transport network in Copenhagen.

Ørestad is located to the south of Copenhagen on the island of Amager and close to Copenhagen international airport. The area was land reclaimed from the sea in the 1930s and 1940s and was owned by the city and national government. The corridor of new development was 5 kilometres in length but just 600 metres wide and led by the Ørestad Development Corporation. The 1995 Masterplan focused new development around a new driverless, elevated light rail system, with two stations, one also connecting with the regional railway. The masterplan aimed for a 20,000 population, 60–80,000 jobs and 20,000 students in the extended University of Copenhagen by 2025, and this has been partly achieved with the exception of the employment target (Knowles, 2012). The new neighbourhood is criticised for its lack of street life and vitality and is less attractive as a residential location than some of the contemporary harbour redevelopments in central Copenhagen. However, the excellent public transport connections and high-quality cycling network, together with car parking mainly being provided in external multi-storey car parks, mean that car usage is relatively low.

**Figure 14.2** The Nørrebrogade and C95 cycle superhighway cycle network, giving convenient and comfortable connections across the city.

## Er det indenfor Cykelafstand?

The resulting culture of cycling can be illustrated by the well-known phrase used in Copenhagen for considering whether a new home or attraction to visit is well located: 'er det indenfor Cykelafstand?', which

**Figure 14.3** The PRACTICE of cycling as the 'modern' and 'progressive' means of travel. Some of the waterfront routes are very attractive, allowing access between the different neighbourhoods.

translates as 'is it within cycleable distance?'. An example new neighbourhood is Sydhavnen, offering a harbourfront setting just a 15-minute cycle ride to Copenhagen central station, which has hence become very popular, in part to its cycling accessibility. The ability to use cycling for

**Figure 14.4**  Lille Langebro cycle bridge, a new connection across the harbour.

travel has become a critical element to liveability in Copenhagen – and the bike is seen as a tool to reverse the damage done to cities over the past century, and a symbol for the newly emerging urban age (Colville-Anderson, 2018).

**Figure 14.5** Cykelslangen bridge, providing an iconic connection to Islands Brygge and the refurbished residential neighbourhood. Cycling facilities are designed with progressive architectural and engineering design, helping cycling to developing a DISCURSIVE MEANING of modernity.

**Figure 14.6** Ørestad neighbourhood. The integration of urban planning and public transport is continued, with a new neighbourhood for 20,000 residents, linked to the city centre by the elevated light rail system.

**Figure 14.7** Residential development in Ørestad. Some of the residential architecture is innovative in design, alongside a high-quality cycle network and cycle parking.

COPENHAGEN   **187**

**Figure 14.8** Communal open space in residential areas. Space that might be given to car parking is used as shared open space, resulting in higher levels of social interaction for children, the elderly and wider members of the new neighbourhoods.

# 15
# Malmö

## Redeveloping the former shipyards

The Bo01 and wider Västra Hamnen (Western Harbour) mixed-use neighbourhoods in Malmö are continuing to be built out. Originally envisaged as the 'City of Tomorrow', low-energy and liveable mixed-use neighbourhoods have been developed, and some of the earliest developments are now over 20 years old (Figure 15.1). The area was originally land reclaimed from the sea and developed as a shipyard and industrial docklands, but the heavy industry declined and was abandoned from the 1980s onwards. The Kockum shipbuilding industry was closed in 1986. SAAB purchased the site and built a modern vehicle manufacturing factory, but this closed as SAAB-Scania merged with General Motors. The land was sold to the city authority in 1996, creating a unique possibility to masterplan at the strategic scale. Unemployment had been high for decades as the industry struggled in Malmö, and redeveloping the Western Harbour provided an opportunity to change the economic profile of the city, but was difficult due to contaminated land and lack of developer interest. The masterplan involved cleaning the industrial sites and rebuilding as a mixed-use neighbourhood, with a planned population of 20,000 residents and 17,000 jobs, including 3 schools, 15 preschools, and an extended University of Malmö for over 25,000 students. Some of the old industrial buildings have been refurbished to draw on the city's distinctive heritage (Malmö City Planning Office, 2015).

## Bo01 residential neighbourhood

Bo01 was one of the first areas to be built in the western part of the Western Harbour, developed as part of the Bo01 Housing Exhibition

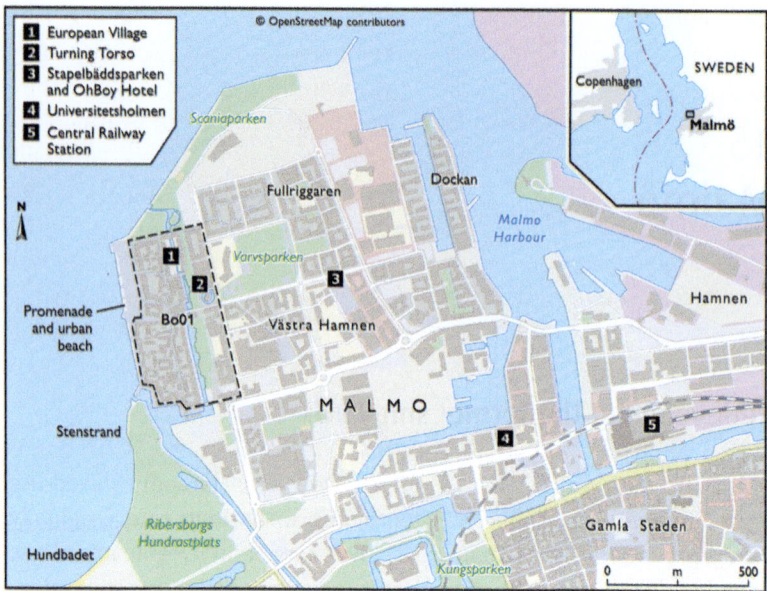

**Figure 15.1** Bo01 and Västra Hamnen.

in 2001, with 'Bo' meaning to dwell in Swedish (Austin, 2013). The area is mainly residential, with some commercial developments such as cafés, restaurants and offices, mostly built at three to five stories in height. The highest residential units overlook the Oresund, the strait of water between Sweden and Denmark, towards the Oresund Bridge and Copenhagen. The higher units also block the winter winds blowing into the rest of the neighbourhood, providing a calmer and warmer microclimate for residents and visitors. The boardwalk and urban beach give spaces overlooking the waterfront, with high usage particularly in the summer, including for swimming in the sea. A winter garden gives a space for sitting during the winter's sunny days. The Turning Torso is the exception in height, a twisted tower with 54 floors, designed by Santiago Calatrava, providing contemporary offices, conference space and residential apartments.

The neighbourhood was masterplanned by Klas Tham to provide a diversity of use, cost and ownership, including rental, private ownership, multi-ownership and student residential homes. The housing styles are purposively varied, with different designs, heights and colours, and a range of architects and developers chosen for building plots. For example, the European Village area includes housing inspired by different European styles, all overlooking the canal. Much of the development area is used for green or open space (up to 50 per cent), providing stormwater management and green roofing, with water integrated into the spaces, together with semi-private courtyards and public spaces. The neighbourhood is carbon neutral; a wind turbine and solar panels provide electricity; and the residential units use renewable energy for heating and cooling, drawing from a district storage system using aquifer thermal energy. There is 100 per cent waste separation and vacuum recycling (Gehl Architects, nd). The development was planned and implemented using a 'Creative Dialogue' between city officials, planners, architects, developers and citizens, enabling debate and knowledge transfer. This can be seen as reflective of a process of communicative action (Habermas, 1984), with participants contributing to the ideas within the masterplan and project development. The dialogues led to the 'Quality Programme', which outlined the expected building and open space standards (Austin, 2013).

## Spaces for social interaction

The spaces in between the buildings are an impressive part of the neighbourhood. The street network and building placements are modified to be less geometric and appear more organic than a formal grid, facilitating a fine network of spaces to walk, sit and dwell, with high-quality landscaping and public realm. The streets give priority to pedestrians and cyclists; through movements are filtered to de-prioritise vehicles; and there are only a few vehicles that attempt to use the residential streets, hence the traffic levels remain very low. A critical design element was to provide space for informal social interaction, particularly in the many semi-private, semi-open landscaped areas. The spaces between buildings are carefully designed, so that pedestrian and cycle movement is prioritised and social interaction is made possible. There is opportunity for meeting neighbours, children can play; and there is passive contact, in that passers-by can be seen and heard. This develops a distinctive discursive formation for transport planning and street design, to encourage walking, cycling and social interaction, drawing on some of the early urban and street design research, which has become more influential in transport planning (such as Whyte, 1980, Gehl, 1987, Jacobs, 1993, Gehl, 2010).

The route to Malmö city centre and the railway station, 1.5 kilometres from Bo01, is undertaken either by bus, cycling or walking. This is one weakness of the masterplan: a tramway connection could have been built; however, there are plans to upgrade the bus route to a tram in future years. Cycling is possible for those with their own bikes, but the cycle and e-scooter hire systems are not so effective, requiring membership of particular private operators and involving high access costs. Car parking outside apartments and car usage within the Western Harbour neighbourhoods are kept to low levels with the use of adjacent multi-storey car parks, often wrapped with retail or residential development to hide the structures. There are also electric vehicle hire and car sharing schemes for those who still wish to use vehicles. The rest of the Western Harbour continues to be built out, including further variety in styles of neighbourhoods, such as the Dockan marina and Flagghusen, aiming at more affordable housing with over 60 per cent of the housing units being rented (Malmö City Planning Office, 2011). Scaniaparken and Varvsparken give open spaces, together with Stapelbäddsparken, an urban skatepark. Alongside, the OhBoy Hotel provides hotel rooms and residential units, with no on-site car parking, apart from one space for disabled people, and cycle parking and hire is provided instead.

## Changing the discursive practice of transport planning

The key innovation from Västra Hamnen is in the diversity of spaces between the buildings, the attention given to street and open space design and the social interaction that this facilitates. The neighbourhood is environmentally sustainable, but there has also been much thought and discussion, using the Creative Dialogues, to implement human-scale environments, positively affecting the experience of living in the area. Transport planning, as a discipline, struggles to measure and value social interaction, instead focusing on traffic counts and metrics such as time saving and congestion. This reinforces the discursive practice of transport planning as a mainly quantitative discipline, attempting to improve the volume and flow of traffic. This practice has now become very outdated, not reflecting the more contemporary public policy objectives required by transport, street planning and urban design. Social interaction, amongst different types of people, is key to vibrant city life. Malmö has successfully achieved high levels of active travel and social interaction in its new neighbourhoods, and it is this end goal that can become a much more important focus in wider transport and urban planning practice.

**Figure 15.2** The Western Harbour redevelopment provides new residential and mixed-use neighbourhoods overlooking the Oresund and on to Denmark.

**Figure 15.3** The 'European Village' area provides different styles of housing overlooking the canal.

**Figure 15.4** The spaces between the buildings are carefully designed and landscaped, so that social interaction is encouraged between residents. This is the DISCURSIVE FORMATION of the Western Harbour.

**Figure 15.5** Streets are designed to restrict and slow traffic, the street network and building placements are modified to be less geometric, and through movements are filtered. A 'Creative Dialogue' was used to develop discussion and debate between city officials, planners, architects, developers and citizens, concerning the content of the masterplan – hence there is a PRACTICE of participatory design.

**Figure 15.6** Secure cycle storage is provided within residential areas.

**Figure 15.7** The Stapelbäddsparken urban skatepark provides community space and the OhBoy Hotel includes bicycles for people staying in the hotel.

**Figure 15.8** Multi-storey car parking is hidden. Some car parking is available, but provided away from residential units and the buildings are wrapped with other uses to conceal the structures.

**Figure 15.9** Central railway station – provides connections across Sweden and on to Copenhagen and other parts of Denmark. Living and working is possible across the region.

# 16
# Dar es Salaam

## Transport for a growing city

Dar es Salaam is the largest city in Tanzania and one of the fastest growing cities in the world, with a population of 4.4 million, projected to grow to 10 million by 2030. The city grew from origins as a coastal fishing village and its name translates as 'home of peace'. A city was first established by Sultan Majid of Muscat in 1857, with a German and British colonial history, and many migrants from India. It was developed into a commercial centre for German East Africa in the late 1800s. The city grew significantly after 1945 and gained independence in 1961. The Tanzanian government followed a socialist path under the Ujamaa (extended family) policies of President Julius Nyerere, but liberalised its economy and governance from 1985 onwards.

Much of the projected urban growth is likely to be unplanned, dispersed and car-dependent if a spatial strategy is not developed and implemented. In particular, there is a critical need for an effective transport system to ensure the city is not overwhelmed by private cars. Buses (68 per cent of trips) are the most common means of transport, followed by walking (17 per cent), private cars (12 per cent) and motorbikes (bodaboda) (2 per cent). The history of public transport in the city dates back to the Ujamaa policies: in 1970 the Dar es Salaam Motor Transport System (DMT) was nationalised and split into Kampuniya Mabasiya Taifa (KAMATA) to operate regional routes and the Usafiri Dar es Salaam (UDA) to operate urban transport (Mkalawa and Haixiao, 2014). Privatisation in the early 2000s led to private bus operators and growth in private cars. The unregulated buses (daladala) included many old, polluting vehicles with unscheduled and unreliable services. Journeys were typically over short distances, but were uncomfortable and slow and the urban poor struggled to pay the costs of travel. Hence, the transport system and travel behaviours were problematised and a more coordinated system planned.

## Improving the quality of public transport with DART

A new bus rapid transit (BRT) system for Dar es Salaam is being developed, named Dar es Salaam BRT (DART). This represents a significant change in transport provision, replacing the informal bus system, attempting to regulate and improve the quality of public transport provision. This follows a number of BRT systems being built in African cities, covering very different urban contexts, including in Lagos, Cape Town and Johannesburg. The BRT systems build on the models of Curitiba's Rede Integrada de Transporte (RIT) and Bogotá's TransMilenio, together with experience latterly found in many cities in China.

DART started with construction from 2012, began operation with its first phase in 2016, and is gradually being extended into a city-wide system (Figure 16.1). The unregulated private buses are being replaced as BRT routes are added. Phase One, running from Morogoro to Kawawa North, Msimbazi Street and Kivukoni Front, had a cost of €134 million and was funded by the African Development Bank, the World Bank and the Government of Tanzania. The route is 21 kilometres in length, connecting outer residential areas to the city centre, with segregated bus lanes on three trunk routes and 29 bus stations. Even in the busy city centre there are segregated routes, which run effectively despite pedestrians using the space. There are 140 buses and more than 160,000 passengers per day, and this is expected to grow to 300 buses and 400,000 passengers per day when fully operational (Magnusson and Kost, 2018).

Four subsequent routes are being planned, eventually leading to 130 kilometres of segregated network and 200 kilometres of feeder services. DART is a major discontinuity in transport strategy terms – improving the quality of public transport provision, including segregated running, express and stopping routes and scheduled services, which dramatically improves accessibility by public transport. Diesel bus vehicles are in operation, though these can be upgraded to electric vehicles in future years. Segregated bus lanes; covered, clean and spacious stations; at-grade pedestrian crossings at bus stations; cycle routes parallel to bus lanes; electronic fare collection; and modern and efficient bus depots are all used (African Development Bank, 2015).

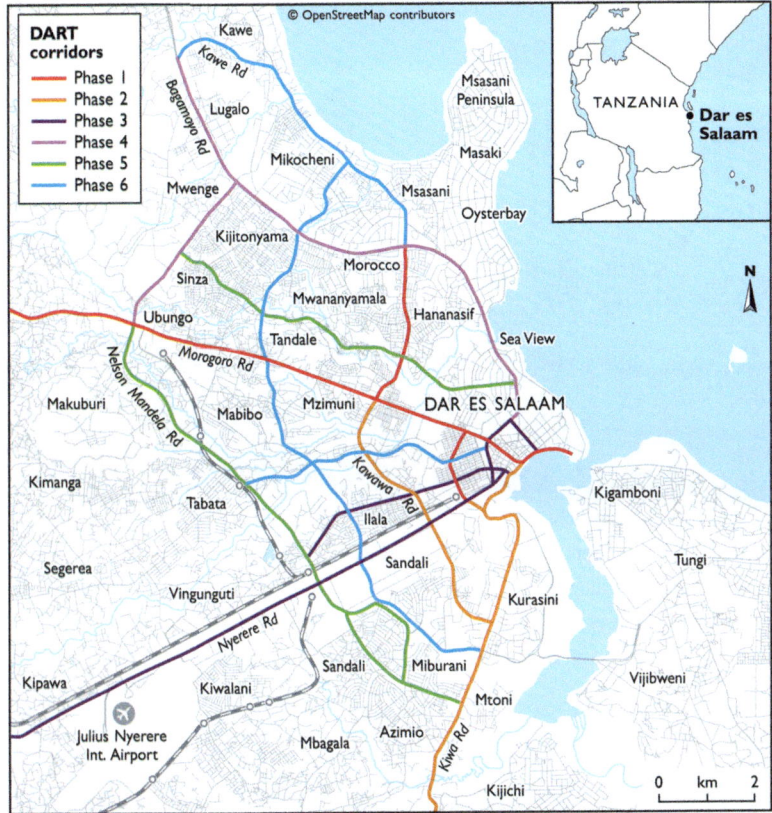

**Figure 16.1** The DART system in Dar es Salaam.

BRT is a necessity in a city such as Dar es Salaam, providing an efficient and inclusive transport system for a growing population. A continued unregulated bus system would struggle to deliver the volumes of people who wish to travel around the city; hence the planning of routes and service frequencies by the city authorities makes for a more efficient public transport system. The BRT system is not without controversy, including the impact on private bus operators, who are invited to participate in shared ownership of DART, but are gaining little financial return and losing from the venture. Hence, there are difficulties in implementation, which can be revised and overcome. But, BRT provision provides a useful model for the development of sustainable urban transport systems in many wider cities.

**Figure 16.2** Emerging mobility in Dar es Salaam, as the city's population increases to 10 million by 2030. It is critical that the projected mobility growth is PROBLEMATISED, so travel behaviours can be modified to be more sustainable in environmental and social outcomes.

**Figure 16.3** The DART system is replacing the unregulated informal bus system and providing a more planned and coordinated public transport system as the revised PRACTICE of public transport.

**Figure 16.4** The modern bus fleet gives improved levels of comfort and the routing and service frequencies are much easier to understand for passengers.

**Figure 16.5** Bus station and waiting environment – giving much improved journey experiences.

**Figure 16.6** Segregated bus routes for DART, hence bus journeys are much quicker than in the congested traffic – this means people will choose to use the bus rapid transit system.

**Figure 16.7** Cycle lanes alongside the bus lanes, giving a safe route for cyclists.

**Figure 16.8** Public transport in the city centre. The vibrant city life is protected as accessibility by public transport is enhanced and more people can travel around the city and participate in activities – this vibrancy becomes the DISCURSIVE MEANING of the transport provision.

# 17
# Delhi

## Regional Metro and regional connections

India is growing in population and urbanising at a rapid pace, with a total population of 1.4 billion, and an urban population of 487 million, which is likely to reach 600 million by 2030 (World Bank, 2022). This urban growth leads to significant demands on infrastructure, including transport. Delhi has a population of around 11 million and the wider National Capital Region (NCR) has 16.8 million. Traffic congestion has been problematic for decades, with vehicular traffic increasing each year. Exact vehicle numbers are not known, but estimates suggest there were 5.8 million registered vehicles in 2010, rising to 11.3 million in 2020, including around 7.8 million two-wheeler scooters (tuk tuks) and 3.3 million cars in 2020 (CEICdata, 2024). Many roads are congested for much of the day and there are competing demands on streetspace, conflicts between vehicles and other users, and high levels of air pollution, dust and noise.

The Delhi-Meerut regional rapid transit system (RRTS) project is a response to these mobility difficulties, building on the success of the Delhi Metro, which opened in 2002 and now extends to 10 lines. The RRTS provides a regional Metro across the city and surrounding NCR, and is the first regional express service in India (Figure 17.1). RRTS aims to improve connectivity between the city and surrounding urban centres, alleviate traffic congestion and promote socio-economic development. The first line of RRTS (phase one) runs from Sarai Kale Khan in Delhi to Ghaziabad and Meerut in the surrounding NCR and State of Uttar Pradesh (Asian Development Bank, 2020). The route length is 82 kilometres, with 24 stations and 2 depots, and runs at speeds of up to 160 kilometres per hour. The route is partly elevated and in tunnel, with a construction cost of ₹300 billion (£2.8 billion) (National Capital

Region Transport Corporation, 2020). The Government of India, City and State authorities and the Asian Development Bank jointly fund the project. The project is partially open (in 2024), running from Sahibabad to Duhai, with the full route expected to be completed in 2025. A typical fare is around ₹20–₹100 (£0.20–£1.00), depending on the length of journey and the transit carriage taken. Ridership is estimated to reach 750,000 passengers per day, rising to over one million passengers per day by 2041. RRTS gives an alternative for journeys made on the National Highway 24 between Delhi and Meerut, with the transit route mostly following the central median of the highway.

The Delhi-Meerut RRTS is the first part of a wider RRTS project that provides regional connections across the NCR (Figure 17.1). Subsequent projects in RRTS (phase one) include routes from Delhi-Sonipat-Panipat and Delhi-Gurugram-Rewari-Alwar. RRTS (phase two) envisages five further routes. The central terminus will be at Sarai Kale Khan in central Delhi, making this an interchange for all of the regional connections. The Ministry of Housing and Urban Affairs is the executing agency for the project and the National Capital Region Transport Corporation (NCRTC) is the implementing agency, acting as a joint sector organisation for the Government of India and State Governments of Delhi, Haryana, Rajasthan and Uttar Pradesh.

## Environmental and social impacts, land value capture

The RRTS project is likely to have very significant environmental and social impacts, beyond those that are conventionally assessed in transport project appraisals. Project appraisal conventionally focuses on time savings relative to cost, using project appraisal guidelines such as those developed by the Asian Development Bank (2017). For example, the Delhi-Meerut RRTS project appraisal (Asian Development Bank, 2020) assumes that there will be mode shift of 15 per cent from two wheelers, 20 per cent from cars and 40 per cent from buses. This leads to quicker journeys, as average speeds on the RRTS are 90 kilometres per hour and on road are 26 kilometres per hour. In addition, road traffic speeds are estimated to increase by 25 per cent post-RRTS, as less vehicles are assumed to be on the road, leading to further time savings. The value of time is assumed to be ₹77.37 per hour. A monetised figure for time savings can hence be calculated annually, reaching over ₹100 million by 2054. There are additional fuel savings, vehicle operating costs and reduced pollution costs. The aggregated benefits are compared to the project cost, giving an

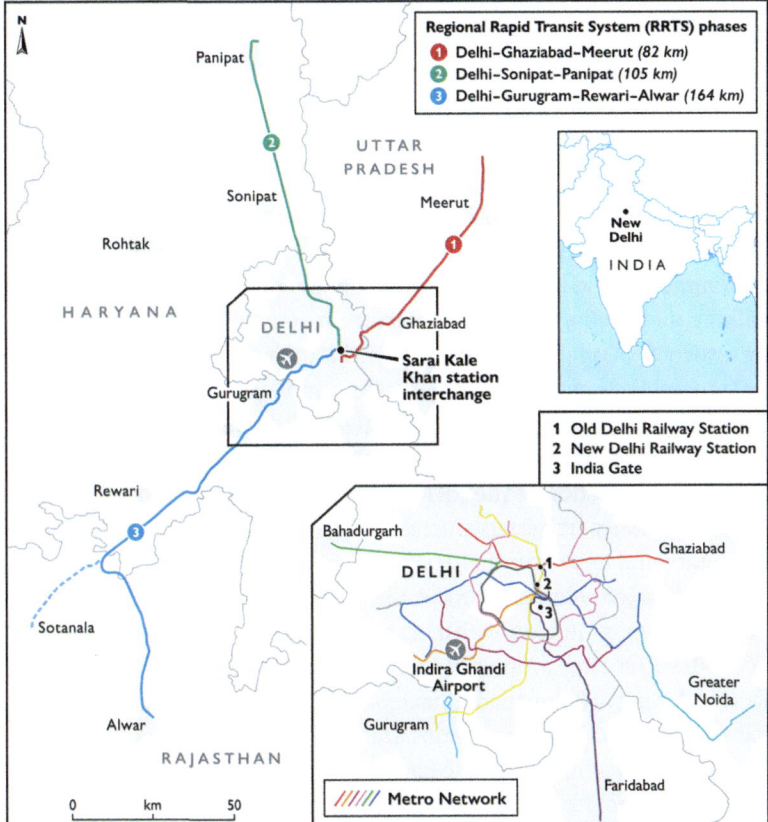

**Figure 17.1** The Delhi Metro, Delhi-Meerut RRTS and additional RRTS (phase one) corridors.

economic internal rate of return of ₹69,744 (11 per cent). Projects with a positive value are deemed worthwhile for society and are prioritised relative to their rate of return and benefit–cost ratio.

This process of economic appraisal is carried out for almost all large transport projects internationally. But, the analysis overlooks the complexity and indirect nature of project impacts (Ackerman and Heinzerling, 2004, Hickman and Dean, 2018). For example, the RRTS will lead to very different travel patterns across the city and region, which are difficult to model, but will be the most significant impacts. Estimated time savings are unlikely to be the most important impacts as there will be a very different set of journeys. The RRTS will facilitate mode shift and also short and longer distance journeys, some of which were not previously possible, with significant environmental impacts in terms of energy consumption, transport $CO_2$ emissions, local air pollution, noise, severance and traffic casualties. There will be significant social impacts as people will be able to access a wider range of employment, leisure, education and wider activities. Urban planning will be important in providing a development form that is shaped around the public transport system, that is, transit-orientated development, and there will be significant land value uplift. Access to RRTS has been kept at a low cost, hence most income groups should be able to access it. The local highway environment could become less traffic-dominated and more pleasant for walking and cycling, but this is reliant on complementary TDM measures, such as taking streetspace away from the private car and reallocating it to public transport (buses), walking and cycling. Hence, the wider transport- and urban-planning strategies will be important for the use of the RRTS. There are additional employment opportunities associated with the construction of the RRTS project and also through working on the public transport services. In addition, there are likely to be wider developmental impacts from RRTS and a transit-orientated development strategy is being adopted to focus new development along the corridor (National Capital Region Transport Corporation, 2024). A catchment of 1,000–1,500 metres around each station is subject to a redevelopment plan, including at Sahibabad, Ghaziabad, Duhai, Meerut, Shatabdi Nagar and Modipuram. Floor area ratios are increased to allow a greater density of development within the catchment areas. Land value capture is being used to help fund the infrastructure, including mechanisms such as land value taxes, betterment levies, development charges or transfer of development rights. Affordable housing is also encouraged, with a target of 30 per cent of residential units being affordable (Ministry of Housing and Urban Affairs, 2017). 'Last mile' transport via bus, tuk tuks, walking and cycling can feasibly increase the catchment of public transport and hence the development opportunities.

Many of these impacts involve the reshaping of the city form and mobility patterns, such is the transformative potential of a regional express public transport system. RRTS will help in moving towards public policy goals, including economic, environmental and social objectives – much beyond the impacts that are measurable through conventional project appraisal. The social mobility aspects are perhaps most interesting, but as yet poorly understood, in that the RRTS will allow a different range of employment and wider activities to be reached in an environmentally acceptable manner. RRTS will lead to much enhanced motility across the city and region, as the range of possibilities for travel and activity participation increase.

**Figure 17.2** Tuk tuk in Delhi. The DISCURSIVE PRACTICE of transport planning has produced a particular form of travel behaviours in Delhi – with the mixing of different vehicles and much congestion and air pollution.

**Figure 17.3** New Delhi Railway Station. Rail travel has a long history in India, stretching back to the late 1800s. New Delhi Railway Station was built in 1956 and is now one of the busiest railway stations in the country, with national services to cities such as Mumbai, Bangalore and Chennai. Direct connections are available on the Delhi Metro and Airport Express.

**Figure 17.4** The Delhi-Meerut RRTS provides the first regional express rail route in India, with connections from the city centre to the wider regional urban centres.

**Figure 17.5** The RRTS station waiting environment provides for comfortable journeys. The RRTS system will be developed into a wider regional network, helping to reshape the city form and mobility patterns.

**Figure 17.6** The low-density city of Delhi is difficult to serve by public transport – 'last mile' connections need to be enhanced to improve pedestrian catchments.

**Figure 17.7** Station area redevelopment opportunities. The RRTS transit-orientated development strategy will assist in providing new residential and mixed-use neighbourhoods within pedestrian or informal transport catchments.

# 18
# Shenzhen

## The integrated transport system

Shenzhen is a fast-growing city, located on the southeast coast of China, in the province of Guangdong. A former fishing village, the city was designated as China's first special economic zone in the 1980s, and has grown rapidly to a population of 17.5 million (2023), forming part of the wider Pearl River Delta urban region. The transformation into a major metropolitan centre has occurred in just 40 years. Urban growth was planned through subsequent city masterplans, developed by the Shenzhen Planning and Land Resource Bureau and Shenzhen City Planning Board. The plans initially focused on economic growth through two development 'belts' and three 'axes', facilitating the development of a major port, information and technology industries and biotechnology, and multiple urban centres (Li and Hu, 2016, UN-Habitat, 2019). Latterly, this has incorporated landscaping, open spaces and an extensive public transport system, as part of a transition towards the 'low carbon city'. The metropolitan area is served by an extensive Metro system, with 16 lines, 373 stations and a route length of 555 kilometres (Figure 18.1) – all of it built within two decades. The first line of the Metro opened in 2004, and now the system is the fifth longest in China and sixth longest in the world. By 2035, there are expected to be 8 express Metro lines and 24 non-express Metro lines and over 1,000 kilometres of route length. There are six mainline railway stations: Shenzhen (Luohu), Shenzhen North, Shenzhen West (Nantou), Shenzhen East (Buji), Futian and Pinghu. Direct high-speed rail connections are available from Shenzhen North and Futian to Hong Kong (Kowloon); and there are further high-speed rail routes to Dongguan, Guangzhou, Wuhan, Shanghai, Beijing, and, indeed, most of the major cities in China. The growth of the metropolitan area and Metro and rail networks reflect a strong emphasis on

**Figure 18.1** Shenzhen's Metro system. Source: based on https://commons.wikimedia.org/wiki/File:Shenzhen_Metro_Linemap.svg. Creative Commons.

integrated urban planning and transport planning, together with rapid project implementation, similar to the strategies adopted in many of the large cities in China over the last 20 years.

## The national transition to electric buses

In addition, and perhaps more well-known, Shenzhen has pioneered the use of electric vehicles (EVs) in its bus fleet. The type of fuel used in public transport fleets is critical to its emission impacts. Diesel-powered buses were once dominant in most city bus fleets in China, but this has changed as the environmental benefits of EVs became more evident and purchase and operational costs were reduced. Electric buses are categorised based on their fuel technologies: pure EVs are powered purely by battery and can include slow-charging or fast-charging vehicles; plug-in hybrid electric vehicles (PHEVs) use hybrid battery and diesel. In 2020, there were around 600,000 electric buses globally, with 585,000 of these in China. The projections are that there will be around 1.7 million electric buses globally in 2030, with most likely to be pure EVs, as government subsidies are largest for these (Deutsche Gesellschaft für Internationale Zusammenarbeit (GIZ), 2018, Eckhouse and Dlouhy, 2021). China remains the global leader for EVs; it has the largest number of low emission vehicles, including pure electric, PHEV and hydrogen fuel cell, utilised in both the passenger car and bus fleets. The manufacturing and use of EVs has risen steadily over the last two decades in China, promoted by central government policies since the early 2000s. This included EV demonstrations at the Beijing Olympic Games, 2008, and Shanghai World Expo, 2010. The Chinese Government's 'Ten Cities, Thousands of Vehicles' demonstration programme, commencing in 2009, signified the first significant step from demonstration to market deployment, using city-based pilots with EVs used in public transport fleets (Li et al., 2016).

## Electric buses in Shenzhen

Shenzhen was one of the pilot cities in China and had the largest EV target. It developed the world's first all-electric bus fleet, using funding originating from the 'Ten Cities, Thousands of Vehicles' programme, together with local government funding from Shenzhen. National and city government investment has been used to promote Shenzhen as the leading, progressive demonstration city running and testing EVs. Different phases of subsidy levels have been used over the years and more than $1 billion is utilised each year to purchase and run 17,000 electric buses and 22,000 electric taxis. Subsidies are usually given direct to the bus manufacturers who subtract these from vehicle sale prices, and recently an operating subsidy has been introduced (Deutsche Gesellschaft für Internationale Zusammenarbeit (GIZ, 2018). EV buses are expensive to purchase, relative to diesel vehicles, costing around 1.8 million yuan (£200,000) per vehicle (Ren, 2018). The subsidy from the Shenzhen city authorities and the national government amounts to around half of the cost of the vehicle, making operations more financially viable. Over 40,000 charging piles have also been built and the leading bus operator, Shenzhen Bus Company, has 180 depots with charging facilities. Buses typically charge overnight and run for an entire day with a driving range of around 200 kilometres per charge (Keegan, 2018). There are a number of challenges to the electrification of bus fleets. The vehicles run relatively short distances before requiring charging, which may be difficult to operate in inter-city transport; initial investment costs for vehicles and charging are high, but operational costs are generally lower than for diesel.

New infrastructure is required for charging and new models of operation, such as the adaptation of services to charging cycles, ranges and electric grid requirements. There are also wider issues that can be difficult to resolve: tailpipe emissions may be reduced, but the lifecycle emissions reduction may be less impressive, depending on the fuel source used for electricity production. More EVs on the streets means that there are more demands for surface-level space. Using mass transit at subsurface-level is a more effective mechanism in terms of providing surface-level space for other uses, but potentially both bus and subsurface mass transit systems can be used in coordination.

## Transferring the Shenzhen experience

Electric bus adoption in Shenzhen illustrates how government subsidy can be used to transition the bus fleet into a more environmentally acceptable form. Again, there is a critical lesson in problematisation, assessing how public transport operations can be most effectively aligned with environmental goals. This has led to significant changes in practice, but required significant governmental support to help fund, test and develop the most effective operational model. The positive environmental impacts of using EVs are demonstrated, with EVs leading to reduced oil consumption, $CO_2$ emissions, local air pollution and noise. Shenzhen has been utilised as a model for EV operation at the city scale, which is now being replicated across China and internationally. There is much interest in the Shenzhen electric bus system to understand how operations can be used and modified elsewhere. Eventually, electric buses should be able to run with much less subsidy and lifecycle costs will be similar to diesel, as running costs are much lower. More than 30 Chinese cities have plans for 100 per cent electrified fleet operations in the next few years, including Guangzhou and Nanjing. Most of the bus vehicles in Shenzhen are manufactured by Shenzhen-based motor manufacturer, BYD, and international sales are becoming important for the Chinese manufacturers. For example, BYD electric buses are now being used in cities such as London. The EV model can also extend beyond the bus fleet, into the taxi fleet, and even into hybrid public–private transport operations and the private car – hence there is huge potential as part of an integrated transport strategy.

**Figure 18.2** High-density city development. The growth in urban population and demand for mobility in large Chinese cities presents unprecedented transport planning problems – requiring PROBLEMATISATION. The planning of Shenzhen is predicated on mass-transit systems and EVs for bus fleets, taxis and private cars. Without these, transport-related air pollution, $CO_2$ emissions and the wider adverse impacts of urban transport become too high.

**Figure 18.3** Mass transit and an EV bus fleet. Shenzhen and the city demonstration programme have demonstrated that EVs can run successfully across extensive bus fleets. This change in PRACTICE requires subsidy as vehicle costs remain high – but demonstrates that EV bus fleet provision is possible at scale.

**Figure 18.4** A pilot city for the EV bus fleet implementation. Many wider cities are following a similar trajectory with EVs, aiming for 100 per cent electric public transit, such as in Guangzhou, Dongguan, Nanjing and Hangzhou.

**Figure 18.5** EVs in the private vehicle market are also being pursued in China, with manufacturers such as BYD becoming more dominant internationally.

# Part VI
# Ethics

## Transport, but for whom?

Ethics is viewed as the relation of the individual or group to others, or even oneself. There is often injustice in human existence and experience, in relation to institutions, practices and individuals, with Foucault arguing that to ignore injustice is to tolerate and perpetuate unjust social practices, thus condemning disadvantaged groups to poor living circumstances (O'Farrell, 2005). Though the terms social justice, equity and equality are often conflated, there are distinctions to be made:

- Social justice: the fair relation between the individual and society, including a 'fair distribution of activities and opportunities within a society' (noun).
- Social equality: the quality or state of being equal (noun), that is, giving the same access to opportunities and resources for all.
- Social equity: the quality or state of being equal (noun) but also taking account of context and circumstance, for example, giving different opportunities and resources to achieve equity in outcome.
- Disadvantaged group: a cohort of the population 'in unfavourable circumstances' (adjective), such as by income, age, gender, ethnicity; and spatially and temporally.
(Hickman et al., 2019, Merriam-Webster Dictionary, 2024)

There are related interpretations of horizontal equity (equality), where individuals and groups are treated the same in the distribution of resources, benefits and costs; and vertical equity, where disadvantaged groups are favoured in order to compensate for inequity, such as through prioritised policy interventions. A key concept is substantive equity, where there is equity of outcomes, via net aggregates of benefits and burdens (Hay, 1995). Experience is a critical part of social equity, defined as the interrelation between types of 'normativity' in a particular culture at a particular time. Understanding issues of exclusion usually involves examining the situation of an individual or group on the 'margins' of society (Foucault, 1975).

In terms of transport-related social equity, the transport systems available allow a specific form of travel and activity participation. By extension, some groups benefit more than others from the transport system as it exists. This is the approach taken to understanding transport's impact on social exclusion, developing from the early 2000s in the UK (Social Exclusion Unit, 2003), which became influential internationally. The analytical focus was to consider how disadvantaged groups may be affected by different levels of transport infrastructure, and hence may experience 'transport disadvantage' (Church et al., 2000, Lucas, 2004, Lucas, 2012). The concept of accessibility was developed, in juxtaposition to mobility, and much of the research considered accessibility levels to particular activities, such as journey time or distance (and wider estimations of utility) to employment, education or health activities. Accessibility was assessed relative to socio-demographics that may also affect inclusion, such as income, deprivation or education levels (Lucas, 2012). This led to the use of accessibility planning in the UK, where new strategies and projects were assessed relative to their impact on accessibility (variously defined) to urban areas, employment and wider facilities. The concept of social capital has also been used to allow analysis relative to individual participation in activities and social networks (Schwanen et al., 2015). The difficulty throughout research in this area remains that social equity is a relational concept and there are no accepted measurements or thresholds in use, indeed they are very difficult to develop. The roles of political and social structures have been underplayed throughout, but are often critical to the existence and use of particular modes and activity participation, such as through reduced state funding for public transport or weak urban planning regimes.

Hence, activity participation is associated with levels and forms of accessibility, but is also more complex, in involving competence (awareness of and ability to make use of the access options available) and appropriation (how the access options are taken up or not, including needs, normativity and aspiration) (Kaufmann et al., 2004). The elements of appropriation are critical in that new infrastructure and accessibility can be provided, but may not be used, for multiple reasons. Recently this debate has been considered using the framework of the Capabilities Approach (Sen, 1985, Sen, 1999, Sen, 2009). Though more often used in development studies, the approach has been applied in transport planning (Nordbakke, 2013, Ryan et al., 2015, Hickman et al., 2017a), with important concepts as follows:

- Capabilities: the alternative combinations of doings and beings that are feasible to achieve, in other words the real opportunities for people to do and to be.
- Functionings: the various things a person may value doing and being, with the realised functionings representing what a person actually achieves and how.

This helps us to differentiate between the 'theoretical' opportunities available to individuals, perhaps relating to new infrastructure provision and changing accessibility levels, relative to what a person may value and wish to do, and actually do (the substantive equity element). Infrastructure provision, including the availability of public transport and other modes, is an important part of activity participation, but there are also wider socio-demographic and cultural issues that may also be critical. Hence, accessibility and activity participation are associated, but modified by other 'conversion factors'. These include individual characteristics, such as income, age, gender, ethnicity, skills and aspiration; the cost of travel; the cultural context and normativity; the shape of the built environment; the governance framework; and more. All of these will influence how well new public transport, walking and cycling facilities are provided and used.

Foucault was interested in how society created institutions and structures, including meanings, knowledge, orders and processes, to help maintain or further social practices. This can involve those experiencing the existing practices, reflecting a particular arrangement of power (O'Farrell, 2005). At the level of the transport system, this requires an examination concerning how society creates and maintains the motorisation-dominant transport systems and high levels of car usage that seem difficult to move beyond. For example, how forms of public transport, walking or cycling facilities are provided, or not provided, in different contexts; how the access costs to public transport are set at high levels; and how dispersed urban developments facilitate the use of the car. There are important differences in interpretations of the individual and the subject. For example, where the individual can be transformed by forms of power into a subject, with a given identity, and controlled to varied extents by others. This is viewed as an exercise of power, including by the institutions producing transport systems, with impacts on different street users. At the scale of the street, this may mean that many are excluded from the use of the street, or certain users are given little space. Colville-Anderson (2018) describes this as the 'arrogance of space', where the street has been almost completely taken over for the use of the car in recent decades, converted into a highway, alongside much space given over to car parking, sometimes with free or inexpensive access costs. The issues are examined with three case studies, each with important linkages between different transport projects and social equity. These are Manchester (transport investment and regeneration), Valenciennes (tramway and social inclusion) and Medellín (transport investment and appropriation of the city).

# 19
# Manchester

## The first modern industrial city and the descent into motorisation

Manchester has a long history of developing the city and its neighbourhoods around transport systems, initially involving canals and railways, and most recently around the Metrolink tramway. The city also has a tradition in political debate, and even in self-promotion, helping to win funds from the UK national government for investment in redevelopment and regeneration. But, alongside, there are concerns that the regeneration initiatives do not significantly turn around the deeply entrenched social deprivation problems found in many of the poorer neighbourhoods in the city and surrounding region (Peck and Ward, 2002).

Manchester is seen as the first modern industrial city, expanding in the early nineteenth century around the cotton manufacturing industry, located at the heart of an international trading network. The first steam-powered cotton mill was built at Shudehill by Richard Arkwright in 1782 and, by the 1840s, coal-powered steam engines drove the cotton mills. This resulted in huge air pollution problems and was associated with very poor, overcrowded living conditions (Engels, 1845, Science and Industry Museum, 2021). The Manchester Ship Canal was built to link the city to Liverpool and the Irish Sea, allowing imports of grain and other materials direct to Manchester from North America, and mostly Canada. A huge engineering feat, the canal was opened by Queen Victoria in 1894, becoming the UK's third largest port, and securing Manchester's economic position for decades. In the 1970s, the docks declined as containerisation took over freight shipping, ships increased in size and the canal system became unnavigable for the larger vessels. The shape of trade also changed, including a closer association with the European market, and the last remaining dock closing in 1982. Railways were also important in the development of Manchester. The Liverpool and Manchester Railway (L&MR) was the first inter-city railway in the world, designed by George Stephenson, opening in 1830 and using steam power. The terminal in Manchester was Liverpool Road, now closed and used as part of the Science and Industry Museum. The network of railways developed across the Manchester region, and indeed the L&MR influenced the development of the railways across the UK from the 1830s onwards.

Greater Manchester now has a population of 2.7 million and this is projected to rise to 3 million by 2040 (Transport for Greater Manchester, 2021). The metropolitan area is still served by a regional rail network, but use of this is constrained by poor connectivity between the northern and southern networks and also the location of railway stations, such as Piccadilly and Victoria stations, which are located on the edge of the city centre. This is a legacy of the separate private railway companies building the early railways and using cheaper land for stations away from the central area (Knowles, 1996). A Piccadilly–Victoria underground link has been proposed since at least the 1970s, but consistently failed to receive funding from the UK national government. Tramways also previously operated in Manchester, from 1877 to 1949, originally horse-drawn and then electric-powered. By 1930, Manchester had the third largest tram system in the UK with a route length of 260 kilometres. But, trolleybuses and rising car ownership led to the closure of the tram system, and trolleybuses were also subsequently withdrawn in 1966. Motorisation grew from the 1950s onwards, including the building of urban highways, such as the Manchester and Salford Inner Relief Route and other radial highway connections.

## Metrolink

The light rail transit system, Metrolink, was a central element of Manchester's urban development from the 1990s onwards. The tramway was originally proposed in 1983 and the first phase opened in 1992. The LRT system mainly utilises former poor-quality suburban railway lines, some disused, which could be converted to light rail at lower cost (Figure 19.1).

Metrolink aimed to improve access to the city centre from the metropolitan area, to assist in the regeneration of the inner urban neighbourhoods and surrounding towns, and also to improve the north–south link across the city by integrating the existing rail networks. Metrolink

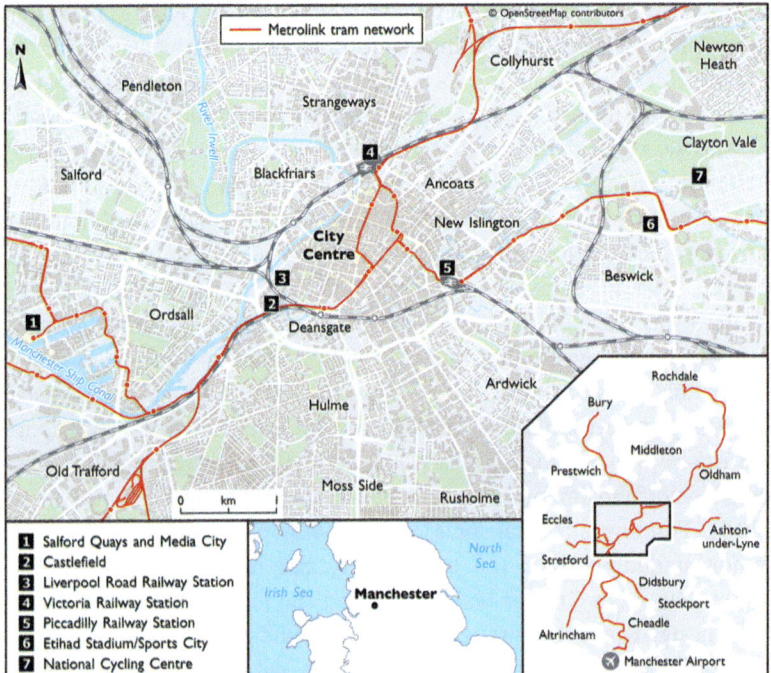

**Figure 19.1** Metrolink and new urban development in Manchester.

was initially owned by the Greater Manchester Passenger Executive (GMPTE), now Transport for Greater Manchester (TfGM), and operated by a private contractor. The operating franchise with Greater Manchester Metro Limited used a design, build, operate and maintain (DBOM) contract. This structure reflected the UK Department for Transport's requirement to include private sector investment in the running of the system and to operate without subsidy (Tyson, 2004). Subsequently, Altram Ltd, Serco, Stagecoach and RATP Group have operated the system. The following stages were used in network implementation:

- Phase 1: Bury to Altrincham via the city centre, including conversion of suburban rail lines and street running in the city centre (31 kilometres route length, opened 1992, with a cost of £145 million, funded by the UK national government).
- Phase 2: Salford Quays and Eccles, a central part of the redevelopment of Salford Quays (6 kilometres route length, opening in 2000–1, with a cost of £160 million, funded by GMPTA, European Regional Development Fund and private developers).

- Phase 3: Manchester Airport and Wythenshawe, East Manchester to Ashton-under-Lyne, Oldham and Rochdale, South Manchester to Chorlton-cum-Hardy and East Didsbury, and an additional 0.4 kilometre spur to Media City and the Lowry Arts Centre in Salford Quays (this was branded as the 'Big Bang' network, 32 kilometres route length, opening in 2009–13, with an initial estimated cost of £489, eventually rising to £820 million, funded partly by the UK national government with £520 million).

Phase 3 was particularly difficult to fund, with the UK national government withdrawing funding in 2004 due to 'rising and excessive' costs. Again, this is a peculiar result of the centralised funding process used for transport investments in the UK; the UK Department for Transport prioritises projects, mainly based on economic assessments, and often refusing to fund important urban public transport projects (Hickman, 2019). Funding approval for Metrolink was reinstated in 2006, with minor amendments to the tram route, and following heavy lobbying by MPs from Greater Manchester and the local media, such as the Manchester Evening News. Complex local funding arrangements were required. GMPTE attempted to raise funds with a congestion charge for Manchester in 2007, but this was rejected in a public referendum. Instead, a Greater Manchester Transport Fund was set up and raised £1.5 billion from a combination of a council tax levy, central government funding, contributions from the Manchester Airports Group, Metrolink fares and the wider private sector. A second tram route through the city centre, passing through Exchange Square, opened in 2014, providing more capacity across the congested central area.

The governmental framework for tramway investment and operation meant it was difficult to raise funding; however, the city has been resourceful in providing funding contributions. Metrolink has been successful in patronage terms, with passenger volumes increasing from 7.6 million per year on the old suburban railway lines in 1992 to 12.1 million per year in 1994 on the Phase 1 route. This was higher than the central forecast and included an estimated 5.8 million trips previously made by the car or new trips relative to 1990 travel patterns (Knowles, 1996). Patronage has grown with an extended network, reaching 19 million trips in 2010 and there were over 45 million trips and a network length of 97 kilometres in 2019. The high-quality tram network offers frequent services (every 6 minutes); with quick, clean, accessible and safe tram vehicles; and gives quicker journey times relative to those by car or bus. The quality of the public spaces around the city centre stations is, however, constrained, with the raised platforms difficult to integrate into the street scene.

This specific history of transport provision in Greater Manchester, with transport systems developed iteratively over time, results in a current mode share for Greater Manchester (all trips) where car and van still dominate (40 per cent) with car passenger (19 per cent); the other means of travel such as pedestrian (27 per cent), bus (1 per cent), Metrolink (1 per cent), rail (1 per cent) and cycle (2 per cent) are still very low in usage (Transport for Greater Manchester, 2017). Hence, despite the recent investments in public transport, there is still a legacy of motorisation and this is difficult to overcome. Manchester's transport strategy (Transport for Greater Manchester, 2021) aims for 50 per cent non-car mode share by 2040 and no 'net increase' in motor vehicle traffic. There are critical social equity dimensions to the transport system that has been developed, seeking to improve access for all, in other words, to connect people to employment and wider activities. There are different forms of public transport, such as the tramway, bus, suburban rail and wider rail links across the network, including a planned High Speed Two connection (but with national funding yet to be approved). Recently, the major networks have been branded as the 'Bee Network', covering public transport, pedestrian and cycle facilities, as an attempt to offer integration across the networks.

## Redevelopment in Castlefield and Salford Quays

The relation of transport investment to urban planning is important to social equity impacts. Metrolink has contributed to the regeneration of many neighbourhoods in the Greater Manchester area. An early example was at Castlefield, the site of the Roman fort, Mamucium, which gave Manchester its name. This is where the Bridgewater canal meets the Rochdale canal, and also the River Irwell and River Medlock. There are four railway viaducts, one for the Manchester South Junction and Altrincham Railway (MSJ&A), the first suburban railway in the city, and also the Cornbrook and Great Northern viaducts and Salford Branch viaduct. Industrial warehouses surround the canal basin, which was designated as a conservation area in 1980 and urban heritage park in 1982. Many of the warehouses have been refurbished into apartments, offices or are used for commercial activities, including Merchant's Warehouse, Grocers' Warehouse and Middle Warehouse (English Heritage, 2002). A pedestrian footbridge, Merchant's Bridge, crosses the basin, giving a well-designed pedestrian crossing across the water. Modern high-rise buildings complement the old warehouse refurbishments, including Beetham Tower and a skyscraper quarter (called 'New Jackson') to the south of Deansgate. Notably, one of the blocks of flats is named 'The Engels', referring back to the mid-1880s of Friedrich Engels. This is an audacious use of the Engels name by the developer, using previous critical histories of capitalism as a sales pitch for expensive apartments. The residents are probably unaware of the cultural dissonance. Similarly, cultural activities have famously been utilised in redevelopment, with popular music (Factory Records, the Haçienda) and classical music (Bridgewater Hall) helping to brand regeneration with a distinctive cultural shape for Manchester.

Following Castlefield, Salford Quays was the focus for regeneration, including using a direct tram connection (Phase 2 of Metrolink) into the city centre. The former docklands, to the west of Manchester, had declined in the 1970s and become a contaminated, derelict industrial area, with many social problems. The area was designated as part of the Salford/Trafford Enterprise Zone in the 1980s, attracting distribution and light industrial businesses. Manchester Docks, the series of docks at the end of the Manchester Ship Canal, including Salford Docks and Pomona Docks, were purchased by Salford City Council using a derelict land grant; a development framework was prepared to guide reclamation and redevelopment. Masterplans were published in 1985 and 1992, with subsequent updates, to allow commercial, residential and leisure uses around the waterfront site. The Lowry Arts Centre was the main focal point, opening in 2000 through central government (Millenium Lottery) and wider funding; it includes two theatres, an art gallery hosting the Lowry Collection and associated cafes and restaurants. Alongside, the Imperial War Museum North was developed and the Lowry footbridge provides a link across the Manchester Ship Canal to Trafford. Media City Metrolink opened in Salford Quays in 1999, with a public transport connection to Manchester city centre, albeit using an indirect and slow route through the Quays. Media City was developed as a mixed-use commercial site by Peel Holdings from 2007, which means that the landowner is making much profit from the development. The main tenants include the BBC, who decentralised some of their TV operations to Manchester; ITV Granada, ITV Studios; Coronation Street production; and the University of Salford. The Quays is now a popular destination for employment and visitor trips, with five million visitors visiting each year (Salford City Council, 2005). There are future plans for the redevelopment of Old Trafford, with a new football ground envisaged for Manchester United.

There are difficult ethical tensions in the development strategy taken at Salford Quays. The approach was similar to many waterfront regeneration projects in the UK and USA, where the private developers were integral to the regeneration process. There are many unresolved issues, including who is gaining from the redevelopment: the local communities do not always benefit from the interventions as might be envisaged. Often the governmental involvement, in this case through regionalising the previously centralised TV production, and providing transport infrastructure, has led to private sector gain as land values rise for sale and rent. There are few effective mechanisms to capture the development gain if the 'regeneration' is framed in this way.

## Emerging new neighbourhoods and transport projects

Ancoats and New Islington provide further examples of regeneration in a difficult economic context. The area was integral to Manchester's industrialisation in the early 1800s, and known for its crowded housing conditions and socio-economic deprivation. The area fell into further decline from the 1960s onwards as the manufacturing industries closed. In the 1950s, over half of Manchester's employment was in manufacturing, but this rapidly declined to less than 20 per cent today. Ancoats typifies the pattern of urban growth, decline and then transformation experienced in Manchester (Peck and Ward, 2002). The area has been comprehensively redeveloped, and parts are still being redeveloped, including the historic cotton mills and warehouses, such as Vulcan Mill, Albion Works, Royal Mill, Beehive Mill (home to the well-known Sankey's Soap nightclub in the 1990s, but now converted to residential use), Ancoats Hospital and Dispensary. Many of the warehouses have been converted into expensive apartments, with a canal-side setting, as part of a new neighbourhood between the Ashton canal and Rochdale canal. Private cars are removed from the central part of the neighbourhood, which is reserved for the canals and pedestrian space. The site is served by a Metrolink line (via New Islington station) with direct links to Manchester city centre and Ashton-under-Lyne. The area is very well located for connections to the city centre, either via Metrolink or a 10-minute walk along the canal-side to Piccadilly and the Northern Quarter, the latter revitalised into an urban neighbourhood with restaurants and bars.

Further up the tram line, the Sports City/Etihad Campus has been developed by the Etihad Group as a sports campus for Manchester City, including the main stadium, academy and women's football stadium and training facilities. The main stadium was originally built for the Commonwealth Games in 2002. The Velodrome, National BMX Centre and National Squash Centre are also located nearby. Lower-income neighbourhoods, such as Droylsden, are also served by the tram line, reflecting the social equity goals of linking deprived neighbourhoods to the city centre employment and wider activities. Metrolink ticket prices are kept relatively low (for example, a £1.40 single fare and £2.70 per day travelcard for Zone 1 and 2). Hence, access to the network is possible for low-income groups, though season tickets, particularly by rail, are prohibitively expensive.

In future years, further infrastructure projects may include tram-trains, which can increase the coverage of the public transport system across the region, such as to Rochdale and Bury, where further deprived neighbourhoods await. A tunnel under Manchester city centre will be required to overcome capacity problems in the central area as more trams use the system. Manchester lies on the High-Speed Two line, and hence could gain quicker connections to Birmingham, London and Glasgow, perhaps in the late 2020s, subject to funding approval from the UK national government. Wider high-speed rail connections (previously branded as Northern Powerhouse Rail) can improve public transport connectivity across the northern region (Transport for Greater Manchester, 2021). These proposals link Manchester to Liverpool, Leeds, Sheffield and Hull, but, again, are dependent on centralised funding approval from the UK Department for Transport. The bus system has been problematic for decades, following deregulation in 1986, which led to poor, competing services and the use of old vehicles. Recently, the bus system has been re-regulated, with Manchester's buses running within a framework reflecting the London system. Through this, TfGM plans the routes, services and fares and private operators bid for franchises to run the routes. Again, there is a particular framing of the governance of transport planning, in this case with engagement of the private sector to operate services, but at least with some regulation of services to move beyond simple profit maximisation. However, strengthened devolution of infrastructure funding from the UK national level to Manchester is critical for the city and the wider region to allow the local governmental agencies to set and deliver their own priorities for future transport infrastructure.

Other areas of transport policy appear more difficult to progress. Pedestrian and cycle facilities and public spaces remain poor relative to the better international practice. There are significant plans for much-needed cycle network improvements: 1,800 miles of cycle routes are envisaged over the next 10 years, at a cost of £1.5 billion (Transport for

Greater Manchester, 2021). Yet, this will not be easy to deliver in areas that are used to using the car for travel – the current levels of cycling are woefully low. Mobility as a Service (MaaS) integration across the public transport services plus electric vehicle charging infrastructure can be further developed, and there are some emerging plans in the current transport strategies. Significant progress is being made towards sustainable urban mobility, and Manchester is often the experimental case study for progressive transport strategies and projects beyond London. But, still, much more can be done beyond the postulation over sustainable travel. Greater Manchester remains car-dependent in the outer areas; some poorer neighbourhoods may have improved accessibility to the centre, but there are difficulties in accessing employment and other activities, in other words, using the new transport networks. The city is particularly poorly linked to neighbouring towns and regions by public transport. Much of the context and governance framework for sustainable transport is unhelpful; for example, public transport remains expensive and cycling and pedestrian provision is very poor. Many in the local population remain averse to changed uses of streetspace.

**Figure 19.2** Deansgate station and surrounding development. Transport infrastructure, including the canals, railways, highways and tramways, has been central to much of Manchester's urban development – a DISCURSIVE FORMATION particular to the city.

**Figure 19.3** Rochdale Canal and Beetham Tower. Contemporary high-rise towers are mixed in with the refurbished warehouses. But there are concerns on the EXPERIENCE of urban life for all, including the inclusivity of the new development. The deep-seated social deprivation is difficult to resolve. Many people remain excluded from the new development and associated activities.

**Figure 19.4** Castlefield Basin and Merchant's Bridge. The revitalisation of Castlefield includes some high-quality pedestrian facilities which improve walkability around the area.

**Figure 19.5** Metrolink in the city centre. The system opened in 1992 and has gradually been extended to provide links across Manchester and the surrounding towns.

**Figure 19.6** Salford Quays and Media City. Phases 2 and 3 of the Metrolink formed an important part of the redevelopment of Salford Quays, with direct links provided to the city centre.

**Figure 19.7** Ancoats and New Islington have also been revitalised as new urban neighbourhoods, with tramway and pedestrian connections into the city centre.

**Figure 19.8** Metrolink in Droylsden. The system provides links to lower-income communities, seeking to improve connections and INCLUSION for residents into the city centre's employment opportunities. But, even where accessibility is improved, there may be wider difficulties with actual usage and activity participation.

**Figure 19.9** Deansgate and cycling provision. There are plans for improvements to cycle infrastructure. But this means of travel has been overlooked in Manchester for decades. It will be difficult to implement high-quality cycling infrastructure and significantly increase cycle mode share from the current low level of usage – investment in cycling facilities remains marginal.

**Figure 19.10** Manchester Piccadilly railway station provides across the UK. But, regional and national rail services are underfunded, such that the cost of rail travel is too expensive for many to use.

## Access to the city for all?

The recent history of urban development and regeneration, coordinated with public transport investment, has turned around a city that was decaying in the 1970s and 1980s. The importance of masterplanning, using urban planning in an integrated manner with transport planning, is well illustrated. Yet, there remain many intractable social deprivation problems. In the latest urban developments, there are concerns of entrepreneurial development, brash gentrification and speculative accumulation through property and middle-class consumption (Peck and Ward, 2002). The deep-seated social equity problems found in some neighbourhoods, including lower educational attainment, crime and poor access to employment and wider activities, are not being resolved. The improved urban development and transport infrastructure do not reach all of the population, or are too costly or difficult to access, particularly from the outer lower-income neighbourhoods. Hence, public transport accessibility has improved, but substantive social inequity remains, particularly across income groups and in specific neighbourhoods. There is a huge problem in appropriation, in that the seemingly improved opportunities are not accessed by all. This demonstrates the complexities in seeking to use transport investment to resolve social equity problems. Even in a progressive city such as Manchester, accessibility levels may be improved, but, still, people may not be able to use the transport networks as envisaged. There is a gap between capability (the doings and beings that are feasible to achieve) and realised functionings (what a person values and actually achieves and how). Hence, there can be much further consideration of what people may value and wish to do, of normativity in different neighbourhoods, relative to what they actually do, and the barriers to participation. This includes transport, as a contributing factor in facilitating activity participation, and the wider conversion factors that affect how people can access and use the city. There remains a lack of understanding of the most effective role that transport can play in improving activity participation. Manchester is often recognised as a progressive city in transport and urban planning terms. But, even here, the challenges for achieving significant progress against environmental and social sustainability goals remain significant for public policy.

# 20
# Valenciennes

## Transport and social development

Valenciennes, located in the northeast of France, provides an example of how public transport investment can help in the redevelopment of an urban area, particularly in a former coal mining area with multiple social deprivation problems. The closure of the textiles, coal and steel industries, in the 1980s and earlier, left a declining population, high levels of unemployment, underused land and neglected infrastructure. There has been a recent revitalisation of the local economy, with growth in the railway industry (including the European Railway Industry, Alstom, Bombardier and other suppliers), automobile production (a Toyota assembly line) and at the University of Valenciennes. Population growth has returned to the city and unemployment has reduced to become lower than the national average.

In 2016, Valenciennes had 45,000 inhabitants and the wider region of 82 municipalities had a population of 400,000. The building of a new tramway, from 2006 onwards, was used to support regeneration and urban development, with the new routes acting as a connection between the city centre, university, new employment activities and the wider regional centres (Figure 20.1). The tramway was developed by SITURV (Syndicat Intercommunal des Transports Urbains de Valenciennes), the city municipality of Valenciennes and the development agency, Valenciennes Métropole. Following the planning of the tramway route in the 1990s, the first section (Line A) opened in 2006, with a length of 9.5 kilometres, serving the city centre, University of Valenciennes and suburban Anzin. An extension to Denain was opened in 2007 (Line B) and a new route between Anzin and Vieux-Condé (Line C) opened in 2014. The tram now has an overall length of 34 kilometres and was built with a cost of €420 million (Hickman and Osborne, 2017).

**Figure 20.1** The Valenciennes tramway.

Line C is innovative in engineering terms as it operates as a single-track, bi-directional system with passing loops; most tramway systems are double-track. The line runs with 15.5 kilometres of single-track and 2.5 kilometres of conventional double-track tramway. The single-track minimises land-take in the urban street and is less expensive for building and operation in a relatively low passenger demand corridor. The Citadis tram vehicles, manufactured by Alstom, with low floors and double doors, allow rapid boarding of passengers, and offer comfortable journey conditions. Each tram, with five vehicles, can hold up to 295 passengers. The route is integrated with the bus system, national rail stops and park and ride. Priority is given to the tram at junctions, and traffic is stopped whilst the tram passes, maintaining route frequencies. Ticket concessions are given to various groups, including job seekers registered with the local unemployment agency who are given free monthly transit passes upon securing a new job. The building of the tramway in Valenciennes, and wider systems in France, are partly funded by the local employment tax, the Versement Transport. This can amount to 1.75 per cent of an employer's payroll, hence giving city authorities a significant income stream that can be used for the development and operation of public transport.

## Public policy beyond economic efficiency

The Valenciennes tramway is different to many tram systems in historic cities: it has been built largely for social objectives. This is a city of relatively low incomes, with a changing economic base and a fairly dispersed population that requires good access to different parts of the urban area and region. The tramway improves the range of activities that can be reached, allowing the local communities to access the new employment opportunities and other activities on offer. The rationale for the tramway is, therefore, different to many urban public transport projects. It is not based on economic efficiency goals, where the project is chosen as the best option of cost relative to benefit, as would occur in more conventional project appraisal processes (Hickman, 2019). Instead, there is a focus on achieving local political goals, which in this case are to regenerate the city and to provide new employment activity for the population.

**Figure 20.2** Valenciennes station tramway stop. The tramway provides high-quality public transport services across the city and wider region.

**Figure 20.3** The tram passing in front of the railway station. There is convenient interchange with the tram and wider national rail network.

**Figure 20.4** Landscaping to reduce the visual impact of the tram. The design and specification of the system is innovative in station and track design, including landscaping, which reduces noise and improves visual amenity.

**Figure 20.5** The single-track bi-directional system minimises land-take in the urban area. The former coal mining area is being revitalised, with the tramway used as a central part of the regeneration initiatives. There are efforts to remove some of the barriers to access with lower fare levels for disadvantaged groups.

**Figure 20.6** Neighbouring towns are made accessible by the tramway – the surrounding areas are linked to the city centre and university campus.

**Figure 20.7** The university is also served by the tramway – as part of seeking to improve access to the city for all. The DISCURSIVE MEANING is that social equity is prioritised in public policy.

This is important in ethical terms, in other words, that social equity is prioritised in public policy. But, also, it leads to a particular approach in the discursive practice of transport planning: the set of rules and apparatus defining project appraisal are refined to allow projects to be assessed against public policy goals. In addition, the difficult issues of appropriation are tackled, so that the improved accessibility is more likely to be used, at least in terms of the cost barriers to using public transport being lowered. There are many important lessons here for wider contexts where project appraisal remains centred on economic efficiency, yet public policy goals require something markedly different.

# 21
# Medellín

## Urban transformation and new urban spaces

Medellín has undergone a radical urban transformation in recent decades following the problematic decades of the 1980s and 1990s, led by a process of civic debate and engagement. The city has a population of 2.4 million and is located in the Aburrá Valley. There has been a longstanding problem with inequity in income and activity participation. The wealthier neighbourhoods are located in the central areas of the city, whilst the informal settlements are found in the periphery, with unplanned neighbourhoods, known locally as barrios, sprawling into the steep mountain sides. These areas accommodate the urban poor, including rural migrants who have flown from civil war in the rural areas. The city has experienced huge social problems relating to the infamous drug baron, Pablo Escobar. His business cartel, at one stage, controlled 60 per cent of the world's cocaine trade. The murder rate in the city reached a global high of 381 murders per 100,000 people in 1991. Escobar was implicated with car bombings, the assassination of presidential candidates and police officers, and violence in the barrios and across the city. Unofficial police militias controlled many neighbourhoods. Economic liberalisation policies exacerbated high rates of unemployment, leading to many willing recruits for the drug trade (Rapid Transition Alliance, 2018).

A crisis point was reached in the scale of violence and the city's elites, including civil society, business leaders and academics, came together to seek to tackle the problems of inequity and exclusion. A series of public debates were held to discuss the crisis and potential solutions, together with community participation in the marginalised neighbourhoods. This urban coalition proposed investments in education, public infrastructure and public spaces across the city, with urban planning as a central feature. The Compromiso Ciudadano (citizen commitment) formed into a political party and Sergio Farjado, a former academic at the Universidad de los Andes and the Universidad Nacional of Colombia, was elected as Mayor from 2003–7, promising to repay the 'historical social debt' in the poorest parts of the city.

This approach to social urbanism introduced a range of urban policies. A new National Constitution in 1991 recognised public space as a constitutional right, allowing new spaces and parks to be justified by politicians. Major investments in urban development and infrastructure began in the early 2000s (Pérez, 2018), leading to an urban transformation that has significantly improved many neighbourhoods and urban spaces. Walking, cycling and social interactions became much safer in many parts of the city. Infrastructure projects included the building of the Metro, tramway, Metroplús (the BRT system) and MetroCable (cable cars), alongside a developing cycle network and Encicla (cycle hire), new public squares, parks and community spaces. Academics were often central to the debates, including Jorge Pérez Jaramillo, Dean of the Faculty of Architecture at Universidad Pontificia Bolivariana (UPB) Medellín in the 1990s, and subsequently Planning Director for Medellín from 2012–5. The city became an urban laboratory, with discussion, debate and citizen participation, including participatory planning and budgeting, where direct participation was used to prioritise funding for projects.

The urban planning process continues, including the Medellín River Masterplan, which provides a strategy for the redevelopment of the area surrounding the river flowing through the centre of the city (Figure 21.1). The new spaces are known as the Parques Del Río (River Parks). The river was canalised in the 1950s and major highways (the Avenida Regional and Autopista Sur) and bridge crossings were built on both sides, utilising space that was deemed available and unimportant to the city. The river and highways became a very unattractive part of the city, creating a huge severance barrier, difficult to access and cross. The riverfront, where accessible, was dangerous and unsafe for visitors, and the river heavily polluted. The redevelopment plans were created by the Metropolitan Urban Plan in 2011, seeking to create high-quality urban spaces along the waterfront, cleaning the river water and bringing the space back into use as a space for outdoor activities. Parques Del Río includes a series of spaces and bridges over the river, in a parkland setting, with the river and the two sides once again connected and used by pedestrians and cyclists.

The first stage of the project has been completed with a new park on the west bank of the Medellín River in the Conquistadores neighbourhood; the highway and other infrastructure has been buried in a tunnel

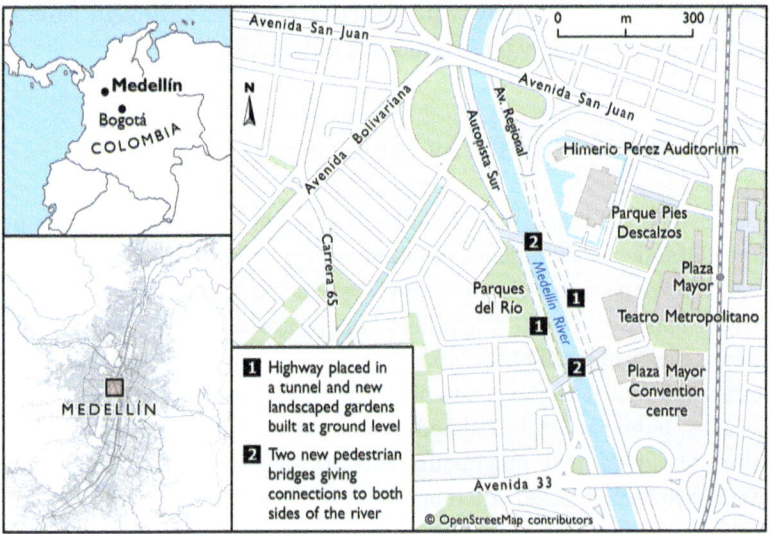

Figure 21.1  New urban spaces along the Medellín River.

with the parkland overhead at ground level. There is a new pedestrian bridge over the river, pedestrian and cycle routes, places to sit and exercise, children's playgrounds, a pavilion café and extensive landscaping. The Teatro Metropolitano de Medellín and Medellín Philharmonic Orchestra are now surrounded by parkland. Twenty kilometres of river bank will eventually be transformed, over 15 years, at a cost of over $1.4 billion (Ortega, 2015).

People will once again be able to use the space along the riverfront, providing a new meeting place for citizens, and a new central park and waterfront for Medellín. The parks will also transform the surrounding neighbourhoods with new residential and mixed-use developments planned. The governance and financing of the project is being carried out by a development company comprising of the mayor's office in partnership with Colombia's two largest utility companies. The company will invest in the redevelopment infrastructure, manage the process, and in return gain real estate development rights, taxes from value uplift and urban taxes. Hence, the publicly owned utility company contributes to urban redevelopment by giving 30 per cent of its profits back to the city. The financial process is inspired by the Porto Maravilha project in Rio de Janeiro which uses a Brazilian pension fund for financing. The Parques Del Río and development in Medellín demonstrate the importance of city planning as a social and political activity – tackling marginality, violence, inaccessibility and inequity, with society transformed through careful debate and high-quality interventions in public spaces.

## Connecting the barrios by cable car

Low-income neighbourhoods tend to be located on the periphery in South American cities, hence urban poverty is associated with problems of access to city centres, lengthy journeys and high costs of travel (Dávila, 2013). This is the case in Medellín, where access from the peripheral barrios has historically been poor, but this has improved with cable car routes linking to the Metro and on to the city centre (Figure 21.2). Over decades, public policy has aimed to improve the quality of life in these informal, unplanned neighbourhoods. Displacement and redevelopment have been rejected and replaced by the gradual upgrading of areas, with improved housing, schools and other social infrastructure (Brand and Dávila, 2011). Enhanced connectivity has been achieved with the MetroCable system and these are particularly suited to the mountainous topography. Lengthy journeys that were previously undertaken by foot or bus, or not at all, are now carried out in minutes by cable car, with easy connection onwards into the city centre by Metro and to employment and other activities.

The idea for the cable car system dates back to an early cable car used for the export of coffee, from the city of Manizales to the Cauca River, to the south of Medellín. MetroCable was the first modern urban cable car system built in South America, and implemented by Mayor Sergio Fajardo, despite some initial local cynicism. Line K of the system connects Comunas 1 and 2 in the northeast to the rest of the city and was the first line with a fixed service schedule. It opened in 2004, carrying 3,000 passengers per hour in the peak, 30,000 people per day, and is connected directly to the Metro system and Metroplús. Construction costs were estimated at $24 million, cheaper than the Metro and BRT, but offering lower capacity. Four cable car lines have subsequently been built, providing connections from different neighbourhoods. In addition, in Comuna 13, one of the poorest communities in Medellín, an electric escalator was built in 2013, taking residents up and down the hillside. Comuna 13 is now visited by many tourists who are interested in seeing life in the barrio, including the colourful graffiti depicting the previous violent times and the local dance and art activities.

The Metro authorities view MetroCable as a socially motivated project, extending the benefits of the Metro to a wider catchment. The

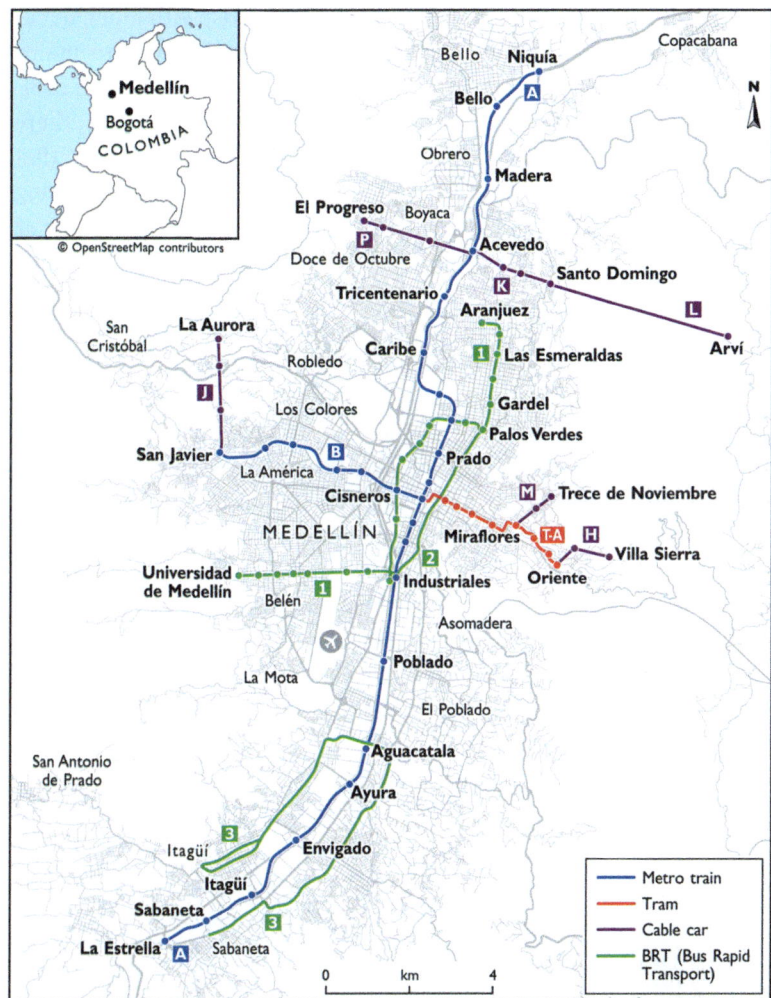

**Figure 21.2** Cable car and integrated Metro and BRT in Medellín.

systems provide highly visible and safer means of travel, with crime and harassment reduced due to the smaller capacity gondolas and camera surveillance. There is often a transformation of neighbourhoods as people can gain access to wider employment activities, with natural surveillance as the gondolas travel over the communities, and there is security presence at interchange stations. This was a major factor in their development – the cable cars allowed the wider population to access previously inaccessible neighbourhoods. There are some perceived problems, including area coverage. Only some neighbourhoods are served; there are capacity issues in peak periods, with much queuing; and the cost of public transport is still unaffordable to many. Indeed, improved access does not mean that local residents can compete in the employment market or participate in other activities; there are wider issues beyond the transport system, often structural, that mitigate against this. There are concerns over the cable cars being used for 'barrio tourism' – with the (often wealthy) tourists acting as voyeurs of the low-income communities. But, overall, this is a very successful system, providing much-needed improvements in accessibility, reducing the lack of mobility that had become an additional form of inequality (Brand and Dávila, 2011). The wider Metroplús BRT system is also innovative in the way that it is run. Different to systems found in Bogotá and elsewhere, the BRT is publicly run. Hence, the city authorities set the framework for bus services and also operate the buses, maintaining more control over operations. Transport planning in Medellín illustrates the benefits of serving an urban planning vision, and directing transport investments for the use of disadvantaged groups, aimed at improving social equity across the city. There has been a major improvement in the quality of life in the barrios and across the city, with the improved transport system providing much-required connectivity.

**Figure 21.3** Overground Metro – providing a north–south movement corridor by public transport, supported by the tramway, Metroplús and MetroCable.

**Figure 21.4** The BRT system (Metroplús) serves suburban neighbourhood centres and the corridors adjacent to the Metro. It is publicly run, allowing city-authority control of operations.

**Figure 21.5** Parques Del Río waterfront renewal – the project brings back the waterfront spaces for use in Medellín. There is greater INCLUSION in use of the public space.

**Figure 21.6** Public spaces for pedestrian use. There are many wider public spaces in the city, allowing social interaction and encouraging urban vitality.

**Figure 21.7** The Encicla system provides public bicycle hire, with stations often located near to the Metro stations and allowing convenient interchanges.

**Figure 21.8** MetroCable serving the hillside informal areas. The system links some of the poorest neighbourhoods to the Metro and enables journeys across the city. The EXPERIENCE of travel is much improved and accessibility is enhanced to employment and activities across the city.

**Figure 21.9** Escalator access to and from Comuna 13, again improving the range of possible mobilities.

## Transport and appropriation

Medellín has experienced a remarkable turnaround in the quality of urban life since the problematic days of the 1980s and 1990s. Urban and transport planning have been central to the process, strongly focused on improving social equity, allowing the opportunity for all to access the city. Public debate and participatory planning have allowed a radical path to be taken, with the Compromiso Ciudadano (citizen commitment) facilitating social urbanism. Redeveloping public space, such as the Parques Del Río, is ongoing, and will lead to much-improved spaces for social interaction. The building of the Metro, MetroCable and Metroplús have facilitated travel around all parts of the city and the range of possible mobilities have much improved. The MetroCable system was particularly innovative and has been replicated in many cities around the world, including in Bogotá and elsewhere, such as in Caracas, Mexico City and Rio de Janeiro. These wider systems are not always so well planned, and illustrate the requirement to directly connect into the wider public transport system, so that interchanges across the public transport system are possible. The focus on ethics is important in demonstrating that all need to be included in city life – and the most effective way of achieving social equity is to improve the lives of the disadvantaged groups. The injustice in human experience has been tackled to a significant degree: in terms of capabilities, the realised functionings of what people actually achieve are much closer to what they value doing and being.

# Part VII
# Subjectivity

## Understanding subjectivities

Achieving more sustainable travel behaviours is often not straightforward, as the most 'effective' strategies, in terms of alignment with environmental, social and economic public policy goals, might not be implemented or even discussed. There may be contestation over the approaches to be taken, including with different key actors, individuals and cohorts within the population. The strategy and projects might be lacking relative to what is possible, and there might not be effective resources to implement. Hence, there can be difficulties in developing sustainable urban mobility strategies at the governmental, societal and individual scales. Even in terms of strategy development, and more evident with the planning of individual projects, there are misplaced assumptions of consensus (Simmie, 1974). Issues of history, power and truth can further conceal the subjectivities that might be evident if only they are able to be revealed.

The concept of subjectivity helps us to understand the likely differences in viewpoints, including the particular experiences or interpretations of reality which differ for each individual and, indeed, governmental or wider organisations. Understanding how these subjectivities relate to transport interventions is critical to project implementation. Foucault (1980, 276) suggests that individuals form their viewpoints and construct themselves in 'an infinite, multiple series of different subjectivities'. Subjectivity can be understood in terms of the regimes of truth, such as how automobility is associated with concepts of 'individualism', 'freedom' and related positive connotations (Böhm et al., 2006), or similar for the different modes. In contemporary, neoliberalist formations, the individual is viewed as self-motivating, acting according to their own views and furthering their individual interests. Actions are often conceived as autonomous and independent from external controls or influence. For example, to drive is normalised as a 'positive', 'modern' lifestyle choice, as a 'necessity' for facilitating busy lives. The adverse impacts on others remain undiscussed or overlooked; driving a car is even presented as your 'right' to choose to drive or to use the streets. The individual, therefore, has been transformed into a subject by forms of power (Faubion, 1994), reflecting control by others and the construction of attachment to particular identities. This position of individual rationality is misleadingly simplistic, as it ignores the structural factors that shape individual subjectivities and behaviours.

There are issues of conduct that are important to understand, including individual and governmental conduct, such as what individuals and organisations do and be. There are rules of conduct that seek to affect individual behaviour, such as the role of institutions and organisations. The role of government can be viewed as the 'conduct of conduct' (Foucault, 1994, 237), but they also act in particular ways themselves. Culture is shaped by the societal or structural influences on conduct, hence can be seen as the 'hierarchical organisation of values, accessible to everybody, but at the same time the occasion of a mechanism of selection and exclusion' (Foucault, 2001, 173).

These issues of subjectivity are examined using two case studies: first, a low traffic neighbourhood in West London (LTN21 West Ealing-South), highlighting the very different viewpoints, at the individual and neighbourhood levels, that can be experienced in relation to a streetscape reallocation project. Over decades, the UK has produced many suburban car-dependent spaces and redesigning these at the neighbourhood level can be more problematic than might be envisaged. Second, a freeway fighting campaign, led by environmental and community activists, is examined for the Interstate-5 highway in Portland, USA. This reflects on the role of the activist campaigns but also the role of city authorities in supporting or disinvesting in particular neighbourhoods and transport systems. As Lefebvre (1974) suggests, space is produced, rather than simply existing; in other words, it is the product of contemporary social structures. It is useful to examine the production of cities, neighbourhoods, streetspaces and transport systems in these terms. Once neighbourhoods and governmental officials have become used to the shape and functioning of their local streetscapes, and the space given to particular modes, the restructuring of these can become difficult and controversial. Discontinuity can be difficult to achieve, even where the current situation may be perceived as unjust, environmentally and/or socially.

# 22
# London, LTN21

## Changing travel behaviours and LTNs

Implementing new transport projects and achieving more sustainable travel behaviours can be controversial and a difficult process – much more than we might envisage as transport planners. Low traffic neighbourhoods (LTNs) have been conceived and implemented in the UK in recent years, initially in London, but also latterly in cities such as Oxford, Bristol, Birmingham and Newcastle. Many LTNs have been successfully implemented, but some provide an illustration of the difficulties found in implementation and provide lessons for wider practice. LTNs are ostensibly designed to reduce vehicular traffic and encourage walking and cycling for short trips in suburban residential areas. The main intervention is to divide residential areas into cells and restrict traffic to travel into and out of one cell, so that vehicular journeys are not allowed across the whole neighbourhood. Traffic is blocked with 'traffic filters', usually in the form of bollards or planter boxes, blocking vehicles travelling through a junction (Living Streets et al., 2020). Walking and cycling are allowed, but the traffic is filtered out. In theory, some of the displaced traffic should disappear, as walking and cycling become quicker and easier to use around the neighbourhood, and these modes are chosen instead of driving. Emergency vehicles are usually exempted from the traffic restrictions. The concept draws on earlier examples of city centre traffic management, such as found in Amsterdam, Delft or Bologna (Pharoah, 1992); however, traffic management with LTNs is applied specifically to suburban residential neighbourhoods. The examples are usually located in the inner suburbs, but feasibly can be used in more dispersed built environments. Waltham Forest provides an earlier precedent in London, attempting to 'Go Dutch' and improve its pedestrian and cycling environment.

The low traffic neighbourhood considered is LTN21 West Ealing-South, a residential neighbourhood located in inner suburban West London (Figure 22.1). This proved to be a controversial project in implementation, at least with some residents, and was ultimately withdrawn as local political leadership conceded to the opposition local resident views (Hickman, 2021, Finn, 2022). LTN21 covers a relatively large area, two kilometres north to south and one kilometre east to west. The neighbourhood is adjacent to Boston Manor and Northfields Underground stations on the Piccadilly Line, two mainline railway stations on the Great Western Railway into Paddington, and numerous bus routes. Hence, the neighbourhood has good public transport connections relative to outer suburban areas in London or beyond. The local cycling network is, however, very poor, with few segregated routes. This includes the connections into Ealing Broadway town centre, which is 2.5 kilometres eastwards (a 30-minute walk and 9-minute cycle ride) and involves cycling in busy traffic.

**Figure 22.1** LTN21 West Ealing-South.

As with other LTNs, the main design principle was to divide the neighbourhood into cells. Traffic access was allowed from the nearest perimeter road into the particular cell, but not through to the other cells; traffic movements through the neighbourhood were restricted using planter boxes. Walking and cycling were allowed throughout the neighbourhood, giving the required 'filtered permeability', meaning that walking and cycling became the most convenient means of travel for many local journeys. The LTN was introduced in August 2020, during the Covid-19 lockdown in London, as a mechanism to quickly implement pedestrian and cycling improvements and to give more streetspace to active travel as traffic levels reduced. Residents were interviewed to assess their views on the LTN (Afonin, 2021), examining the language of those for and against the project. The analysis draws on the approach of content analysis from Fairclough (1992a) and the framework of Lefebvre's (1974) spatial triad. The latter framework helps to highlight the differences between conceived space (the 'official' representation of space), relative to perceived space (the understanding and use of space by residents) and lived space (the experience and value of space) (Hickman and Afonin, 2022, 2025). The premise is that there are different views and experiences of the new streetspace arrangements and that these should be more fully understood and responded to in the planning and design of streetscape projects.

## Conceived, perceived and lived space

Conceived space can be viewed as the governmental representation of space, such as what was planned (Lefebvre, 1974). In this case, this is largely the narrative given by the transport planners, urban planners and politicians at Ealing Council, covering the objectives for the LTN and the design of the project. The stated objectives of the LTN were to reduce through-traffic and to improve the environment for walking and cycling, allowing more people to choose these modes rather than the private car or taxi (Ealing Council, 2021). Even the terminology is difficult here: the pejorative framing of through-traffic using the metaphor of 'rat running' gives prominence to the movements through the area, raises objections from car drivers who dislike being labelled in this way, and downplays the resident-originating traffic that also needs to reduce. Through-traffic in the residential neighbourhood is actually likely to be a low proportion of traffic movements and much of the circulating traffic is probably from residents or visitors to the neighbourhood. However, there is no available traffic movement data to substantiate these claims. The LTN was introduced through an Experimental Traffic Order (ETO), which allowed a quick installation of measures. This, again, was viewed positively by the borough council, as a mechanism to introduce measures quickly. Residents were informed of the LTN only a week in advance of implementation, as required with an ETO. The project was carried out on an experimental basis, initially envisaged for six months, with residents and others able to respond to a statutory consultation at the end of this period. Hence, the idea was to implement quickly and in an experimental manner, so that residents would see the project in action, and presumably view it more favourably after a few months.

Perceived space can be seen as the understanding and use of space by residents and others (Lefebvre, 1974). There were many differing views on the LTN, including those from local residents, employees, Ealing Cycling Campaign and resident groups set up to oppose or support the LTN, such as One Ealing, Ealing Better Streets and Coldershaw and Midhurst Traffic Action Group (CAMTAG). Generally, the supporters of the LTN and new streetscape layout enjoyed the reduced traffic volumes and air pollution, the quiet streets, and the improved conditions for pedestrians, cyclists and children playing out in the streets. There were many more people walking and cycling through the neighbourhood.

> I think they [LTNs] have opened peoples' eyes to the fact that they could be living in a much more pleasant, calm, lovely environment than they were. Once it was embedded, they found their roads were blissfully quiet and there weren't cars from miles away doing speed runs down the road, it was actually incredibly pleasant. We should be demanding nicer environments to live in safety with good health – lots of trees, lots of planting, lots of open space – access to water and fresh air, good facilities, safety, quiet, peace. We should be allowed to have these things in a modernised country. Why wouldn't you want more of that – everyone benefits? (Interview 5)

The residents against the LTN were concerned about longer journeys and difficulties in accessing activities across the borough, such as health appointments and employment or leisure trips. There were concerns over displaced traffic and congestion on the adjacent roads, that the consultation was ineffective and that the project had been 'imposed' on the population. There is a strong reflection of condition, in that individual circumstances are reflected in the views given.

> I was forced to drive down Boston Road and it created a lot of congestion very often back down to my own road and up to Uxbridge Road, which is gone now that they've removed the LTN. It was causing the buses not being able to get along, so their schedule must have been all mucked up. The ambulances couldn't get down, because Boston Road leads to Uxbridge Road near the hospital, so the ambulances were getting held up. (Interview 1)
>
> I noticed that my hospital visits started getting longer and longer and that my 65 route, which is an absolutely vital key commuter bus route, as well as for old people to get to hospital, started coming up there with signs saying we don't know when the next one is coming. It could be 52 minutes, there was just no information, and I was late for a couple of really important appointments that I left a lot of time for. (Interview 2)
>
> Every time I go out of my house, instead of having a three-minute journey to head towards the A40 and Pitshanger and that area, I now have to travel an extra 20–25 minutes, depending on the traffic, to go all around Ealing. I have to do that at least once a day, if not multiple times a day. It's my route to the people I work with, it's my route to my GP, it's my route to the supermarket, it's my route to the A40, it's basically my route to most of the places to which I go. (Interview 6)

The more stringent opponents argued that more streetspace was not needed for pedestrians and cyclists, and that children would not want to use the new space; even that the project was implemented primarily to increase local authority revenue via traffic fines, and that the quiet streets had become too quiet and unsafe for pedestrians, particularly for women. Hence, there is often a lack of coherence in argument from the opponents, such as that Ealing is already 'green' with many open spaces and there is no need to move towards more sustainable travel behaviours. Views reflect particular worldviews and social constructions; people see themselves as 'victims' of the local authority's LTN interventions.

> There was never an issue in Ealing, ever. So, they already did everything they implemented. We've got huge green spaces, we plant trees, we have cycles lanes and there are many people who only drive out of necessity and they shouldn't be fined for doing so and I think they did everything we could. (Interview 2)
>
> I pay my taxes to support all of the roads in this area, why should I be banned from driving down some of them? If they want to turn

> them into private estates that's fine – the people who live on those roads can play on those roads, but I pay a tax for the whole borough. They're not needed, there's no point to them. We don't want people walking in the middle of the road, we don't want kids walking in the middle of the road. It's just absurd. (Interview 4)

Lastly, lived space includes the representational meaning, including the experience, value and depiction of space by residents and others (Lefebvre, 1974). It covers the emotions of using space, and can include resistance to the given space. There were varied positive and negative reactions to the LTN and the space produced. For example, the LTN supporters changed the 'road closure' signs to show 'road open for pedestrians and cyclists', highlighting the importance of language even in the signage used for the project. Some residents attempted to collect data, in the absence of local authority data, to show how traffic volumes had decreased and the numbers of pedestrians and cyclists had increased. Some of the antagonists protested at local authority offices, and even vandalised the planters and poured oil on the cycle routes to deter the cyclists. There is, again, use of metaphor to illustrate the emotions being felt, such as the LTN leading to living in a 'road prison'.

> When the LTN was first implemented, many of us predicted what the effect would be, and we were proved correct, but the Council absolutely refuses to believe it. They are adamant that it's all wonderful and does all the things it is expected to do, makes people walk and reduces pollution, which it doesn't at all. And, incidentally, they are also terribly keen on cycling and all we have now is cyclists riding on pavements and terrifying the pedestrians half of the time. (Interview 1)

> I just thought, am I living in a democracy or am I living in war-torn Russia? I thought this is outrageous, this is our freedom and I can't see my friends, they can't come visit me because of these bloody great concrete slabs …. So, I wrote to Ealing Council, wrote to my MP, wrote to BBC News, and said what is this thing that's going on …. Now I feel like I'm in a road prison and I can't get anywhere. (Interview 2)

Ultimately, the LTN project became very controversial with a minority of residents, who became very vocal. The project was subject to legal challenge by a group of residents. Ealing Council issued new LTN Orders in February 2021, revising restrictions on emergency services and for disabled residents. The consultation period was extended to August 2021. In May 2021, the LTN was a major factor in the local council election. There was a change of leadership at Ealing Council, with the Leader of the Council, Julian Bell (Labour) replaced through the election. The new Leader of the Council, Peter Mason (also Labour), won by promising 'to listen' to the residents. He stated that

> The Council will be transparent and inclusive ... that the people will be given control over change in their neighbourhoods.

To understand support for the LTN, a survey was undertaken by the local authority, using SurveyMonkey software, resulting in 22,000 responses. Four thousand respondents reported that they lived in the LTN and 1,000 on the adjacent roads; 17,000 were non-LTN residents. This was an unsampled survey with a very low response rate (6 per cent) relative to the 345,000 residents in the borough, and it was not known exactly where respondents were located. The survey asked 'Would you like the LTN to be made permanent once the trial period ends?' 82 per cent of respondents said 'No' (Steer, 2021). On this insubstantial basis, the LTN trial was abandoned by Ealing Council and the barriers and planters were removed. As a result, the cars and traffic have been returned to the neighbourhood streets.

## Overlooking car usage in the suburbs?

The history of LTN project implementation, in this case of LTN21, involved a well-intentioned project. It aimed to reduce traffic levels in part of inner suburban London, but was poorly implemented, with very poor participatory procedures. The regime of truth shaped the discourse that became accepted, in other words, that the project was too controversial to be continued. Within this, there were many supporters who wished to see the project develop, despite its poor design, but a significant cohort of opponents vehemently disliked the project. The problem is that this type of neighbourhood experiences traffic levels that are too high and travel behaviours that are environmentally and socially unsustainable. As Lefebvre (1974) suggests, decades of capitalism have affected the perception and use of streetspace. Streets have been given over to the car, and pedestrians and cyclists have lost the opportunity to use the street. In the main, this remains unquestioned. The motor manufacturers and associated organisations have persuaded many people that

**Figure 22.2** Wooden planters to block through traffic movements. LTN21 West Ealing-South involves relatively marginal interventions, the wooden planters acting to filter traffic so that car driving becomes less convenient and walking and cycling are preferred.

**Figure 22.3** Improved walking and cycling environments. There is support for these types of streetspace redesign projects, as the walking and cycling environments are much improved; most people support an improved residential neighbourhood with less traffic. Source: Jack Fifield, Flickr, Creative Commons CC BY- SA 2.0.

**Figure 22.4** Opposition to LTN21. There can be some vocal opposition from a minority of residents. The opposing residents produce a REGIME OF TRUTH that conflates the use of the car with a defence of their freedoms, rooted in their own perceptions of the world and reality. This requires very careful debate and mediation. Source: Roger Green, Flickr, Creative Commons CC BY- SA 2.0.

they should travel around in their vehicles (as an expensive product to be purchased and consumed), and urban planners and traffic engineers have shaped the built environment accordingly over decades. Many residents have become used to this way of life, and the car is embedded in and facilitates their lives. The car has been normalised as the way that most people access activities, from employment to retail, heath and leisure. Of course, there are different views, and there are many supporters of public transport, walking and cycling – but still the resources that are spent on motorisation massively overshadow the resources for public transport, walking and cycling. The regime of truth of motorisation has become generally sanctioned, and the actions of transport planners and politicians are not effective, including in this LTN case, in shaping a different reality. Indeed, the techniques and procedures of transport planning are helping to shape what is viewed as true.

The hegemony of motorisation is in two forms. First, in terms of governance, central government removes itself from the practice of any form of significant interventions at the neighbourhood and streetscape scales, while investing billions in increased strategic highway capacity

**Figure 22.5** LTN21 is removed and the car returns to dominate the street – and little progress has been made on sustainable urban mobility. The CULTURE of the neighbourhood represents the hierarchical organisation of values as a mechanism of selection and exclusion. In this case, residents without access to a car are excluded from a living environment that is traffic-free and struggle to participate in many activities.

and making motoring cheaper through reduced taxes. Local authorities are asked to implement sustainable travel initiatives at the neighbourhood level, but given little in resources to implement projects carefully and robustly. There is insufficient funding provided for wider public transport, including for buses; there are few orbital public transport connections; cycling facilities in Outer London are very poor; and cycling remains unsafe for the vast majority of people. There is much inconsistency in policy approach, in the government's conduct of conduct, including purposeful misrepresentation. For example, suggesting climate change and social equity are important policy objectives, but doing little to tackle them.

Second, a vocal minority of the population strongly believes that the car is crucial in their lives, enabling their participation in activities and life. They perceive that car access and usage are matters of right for the individual, irrespective of societal impact, of which they are unaware or do not care. They conflate the use of the car with a defence of their

**Figure 22.6** Richmond Park and cycling during the Covid-19 lockdown. Across London, many of the residential neighbourhoods, and even some of the larger Royal Parks and open spaces, remain traffic-dominated, despite some some recent efforts to reallocate streetspace. There is huge potential to give much more space to walking and cycling – but the aspiration to do this needs to be much more ambitious.
Source: Jack Fifield, Flickr, Creative Commons CC BY- SA 2.0.

freedoms, rooted in their own perceptions of the world and reality. The use of the car, in the suburbs, is perceived, by many, as what 'normal' people do. One powerful cohort, in this case car owners and drivers, are using space for their own ends, at the expense of other users of space. Pedestrians and cyclists, the young and the old, who have no access or only partial access to the car, are marginalised from use of the street. This is the localised culture and normativity as the organisation of values and mechanism of selection and exclusion. The societal context for the production and use of streetspace in LTNs incorporates social actions, individual and collective, including expression and suppression. Ask a child in West Ealing if they would like to use the streetspace in a different way, and they will be likely to suggest some more playspace or safe cycling routes – but, of course, they remain unheard during the transport-planning process. Other LTNs have been more successful and the support for streetspace interventions has been maintained and offset any objections. This is an example of Bourdieu's (Hardy, 2014) hysteresis, where a change in field structure (policy changes towards sustainable mobility) have led to a mismatch in habitus for some (the opponents perceive that they cannot accommodate the change in priority given to the car). The general lesson is that a much stronger participatory and deliberative process is required to implement this type of project, otherwise a vocal minority can hijack the process and difficulties follow in implementation. It will not be possible to reach consensus for all, but there needs to be demonstrable support to offset any objections, and this can only be achieved through increased deliberation on the project through project planning and implementation.

# 23
# Portland

## Urban development and inclusion

Portland, located in the state of Oregon, USA, has an urban population of around 650,000, with 2.5 million in the wider metropolitan region in 2020 (United States Census Bureau, 2020). Portland was originally a port city, trading in timber, but, from the 1960s, developed a reputation for liberal and progressive political values, and became a centre for music and counterculture. The city is often put forward as an exemplar of sustainable urban development in the USA, known as the 'Portland model'. The recent history of urban planning has produced urban densification, urban growth containment, a light rail system, walkable neighbourhoods and a cycle network in the city centre, extending into the inner suburbs. There is a compact urban centre and greater availability of environmentally sustainable transport modes than virtually all cities in the USA. Mode share in the Portland urban area is less private car-orientated than the rest of the USA, with car and car passenger/car share at around 67 per cent of commute trips, public transport at 11 per cent, work at home at 8 per cent, walk at 7 per cent and cycle at 6 per cent in 2012 (City of Portland and Multnomah County, 2015). The metropolitan region is more car-dependent with cars and car passengers at around 90 per cent of trips (City of Portland, 2012, Walker, 2010). This data predates the Covid-19 period, when travel behaviours are likely to have changed, including much more working at home; feasibly, trips into the city centre are less prevalent and the commute may be less important relative to other trip types. A high level of car dependency will remain, particularly for trips outside the inner urban area. There is progress towards sustainable urban mobility relative to the rest of the USA, yet this is relative and does not compare well to

European practice. There remains much highway infrastructure provision within and surrounding the city, alongside the emerging light rail network (Figure 23.1).

In addition, there have been critical concerns, over decades, relating to the inclusivity of the new urban development and the environmentally sustainable city lifestyles that are produced (Goodling et al., 2015). This includes a problematic history of institutionally racist planning and development practices, with the emergence of segregated low-income neighbourhoods, such as Albina, which has been home to a predominantly Black population (from the 1950s and 1960s) (Gibson, 2007). The area was subject to subsequent redevelopment to facilitate urban freeways, including the Interstate-5 (I-5) highway (1970s). The governance of urban development and the I-5 highway project is examined along with the subjectivities that arise, including environmental and community activism against highway expansion plans.

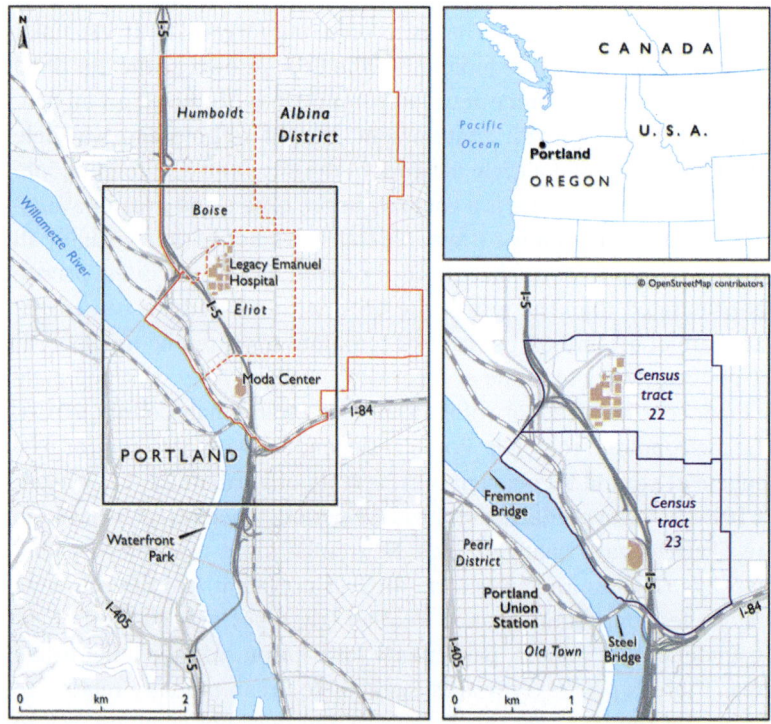

**Figure 23.1** Portland and highway infrastructure provision.

# Interstate highway investment and freeway removal projects

As is well-known, the USA has been at the forefront of motorisation and has prioritised the use of the private car through transport and city planning for more than one hundred years (Norton, 2011). An extensive network of urban freeways and interstate highways has been developed, the most extensive in the world, alongside dispersed metropolitan areas, with continued and significant highway investment. Highway planning was initially promoted through the Federal Aid Road Act (1916) providing the first Federal funding for highways and initiating the building of a national highway network. This included highway design features that prioritised car usage, such as controlled access, junction design and a range of unified engineering design standards. Movement of the private car through and between cities was prioritised by giving extensive street and parking space to the car. The principle of free access for motorists (in other words, free use of the highways, low taxation on gasoline) was applied, which became an important feature of mass motorisation. This contrasts with the often-expensive access to public transport (in other words, costly public transport fares) and insufficient networks. President Dwight Eisenhower's National Interstate and Defense Highways Act (1956) provided significant funding and coordination, with $25 billion (equating to $215 billion in 2023) provided for 66,000 kilometres of interstate highways, to be built over 10 years. This was a transfer of funds from the National Defense budget and represented the largest public works investment at the time for the USA (Rose and Rohl, 2012). The Highway Trust Fund has since provided funds for extending and maintaining the highway system, derived from gasoline tax, with 90 per cent of funding for new highway projects coming from the Federal level and 10 per cent from State level. The interstate highway system now extends to 78,000 kilometres. Yet the extent of motorisation means that the interstate highway system only accounts for around 1 per cent of highways and 25 per cent of vehicle kilometres travelled (VKT) across the USA (US Department of Transportation, 2021b, US Department of Transportation, 2021a). The promotion of motorisation as the primary means of travel continues with further projects to extend highway systems, serving increasingly sprawling cities. In governmentality terms, there is a very distinct approach taken to transport planning, developed at the Federal and State levels, with heavy prioritisation of funding for highways at the expense of other modes. There is a 'conduct of conduct', which produces particular travel behaviours, usually with significant dependency on the car, even where there are limited attempts to introduce public transport. Despite massive highway provision – 10–15 lane highways in each direction are not uncommon in some cities – major parts of the highway system remain inadequate relative to traffic demand and are becoming old and expensive to maintain.

The case for more highway provision is produced across the USA, usually by the Federal and State authorities, sometimes in accord with the city authorities. Support is given by motor manufacturers and associated industries, with projects planned and implemented by the transport-planning and urban-planning consultancies. The motorisation approach is often supported by politicians and residents, with highways deemed as critical to city growth or individual mobility. Many narratives have been used to support increased motorisation. From the 1950s onwards, urban expressways have been perceived as a means to rescue struggling city centre economies and to bring the 'white flight' suburban dwellers back into the city centres. There is little critical reflection that compact urban development may necessitate less highway provision, that highway capacity induces traffic demand, or of the adverse environmental and social impacts that follow motorisation. At the individual level, more and more time is spent travelling from home to destination activities. The social equity dimensions of highway provision are often significant, with neighbouring communities adversely impacted by highway severance, noise, poor air quality and displacement. Existing neighbourhoods have been demolished to provide space for highway provision and wider 'urban renewal' projects. This led to much debate on the so-called comprehensive masterplanning approach (Jacobs, 1961, Mumford, 1963). The interstate highway programme was a Federal effort, produced for an assumed national 'modernism', impacting on local communities and with little public participation: 37,000 housing units were demolished for highway building and residents displaced in the USA from 1956–66 (Avila, 2014). Transport planners and engineers understood their narrow remit as the building of highways, overlooking any impacts on communities or, indeed, the wider environmental and social impacts. There were disproportionate impacts on lower-income neighbourhoods, often for Black, Hispanic and wider communities of colour. Hence, major social equity impacts were experienced in facilitating motorisation, yet these were overlooked by transport and city planners. It was assumed that distributional issues were of little importance for transport planning. In addition, those without access to a car remain unable to access many activities which have spread across the dispersed city and public transport connections are so lacking that activities cannot be reached, including employment and wider activities.

Yet, in some cities, there are community activists offering more inclusive possibilities for transport planning. There is contestation over the highway-focused transport-planning approach, highlighting the community and environmental impacts of urban freeway development. A freeway removal movement has emerged, with origins from the 1970s, and gradually gaining influence, as citizen groups, politicians and transport planners realise that there are major community, environmental and safety benefits to using highways and streetspace in a different manner. Highway expansion projects have been blocked and alternative plans are envisaged, for example, removing, burying or decking over parts of the highway networks, and using the surface-level space for urban development, open spaces and improved provision for public transport, walking and cycling. Neighbourhoods previously severed by highways are sometimes reconnected through the removal process. There can be quite widespread support for these interventions. There are significant possibilities for urban redevelopment, including for new commercial and residential development, entertainment and cultural districts and public spaces. There are different types of freeway removal projects, some removing or replacing highway capacity (for example, in tunnel or decked highways), with associated public transport provision, and redevelopment plans for the new spaces. Interventions are seen, by some activists, as the emerging, contemporary response to the failures of the USA-inspired highway planning approach (Rose and Rohl, 2012). Some high-profile freeway removal projects have been implemented over recent decades in the USA, following the removal of Harbor Drive in Portland as the first significant freeway removal project in the country. The highway was constructed in the 1940s, running adjacent to the downtown, along the western bank of the Willamette River, providing three lanes in each direction as part of Route 99W. This became less frequently used as further urban highways, such as the I-5 and I-405, were built. In the 1960s, in the midst of growing freeway revolts in many cities and an emerging environmental movement, the Riverfront for People citizens group called for Harbor Drive to be removed and a riverfront park to be built. In the late 1960s, Governor Tom McCall, a former environmental activist and journalist, promoted plans for the replacement city park. Particularly, he viewed this as part of the response to industrial decline in the city. The ideas were inspired by the historic, unimplemented masterplans for Portland, reflecting 'City Beautiful' principles, dating back to the early 1900s. Following much debate, the freeway was removed in 1974 and replaced with a riverfront park in 1978 (McVoy, 1945, Congress for the New Urbanism, 2009). The park was latterly renamed as the Tom McCall Waterfront Park. The removal of the freeway was seen as an important contributor to the subsequent revitalisation of Portland's downtown area, becoming emblematic of urban renewal efforts in the USA. The new open spaces were supported by major new urban infill developments to the north and south, such as The Yards at Union Station and RiverPlace.

Across the USA, further freeways were removed or partially blocked, in cities such as Baltimore (the I-170 Franklin-Mulberry Expressway, 1981); San Francisco (Embarcadero Freeway and Central Freeway, opening 1991, 2002); New York (West Side Elevated Highway, 2001); Chattanooga (Riverfront Parkway, 2004); and Boston (Central Artery of I-93, known as 'The Big Dig', 2003). The Boston Big Dig additionally became known for its lengthy and overrunning implementation schedule and project cost, but was more successful in urban-planning terms by reducing the highway severance in the central city area. Freeway removal projects, typically, take decades in debate and contestation, before plans are agreed and implemented. But, of course, motorisation remains heavily dominant in most cities in the USA, and the dispersed nature of the built form means that investing in public transport networks is costly and gaining sufficient patronage within sparse walking catchments is difficult. Some freeways have been blocked or removed, but largely the Highway Trust Fund remained intact.

## The I-5 freeway and Albina

Alongside the history of freeway removal in Portland, the development of the I-5 highway through low-income neighbourhoods, such as Albina, was very controversial. The I-5 is the major north–south route along the west coast of the USA, stretching from Seattle to San Diego, hence viewed as an important interstate route. In Portland, the I-5 runs along the eastern side of the Willamette River, overlooking the city centre and Waterfront Park, travelling northwards into the Albina neighbourhood. In the 1950s–60s, Albina was predominantly a neighbourhood for a Black population, reflecting segregation in the city, and the result of institutional racism from the city authorities, developers and mortgage brokers. This included disinvestment in the housing stock, lack of access to mortgage lending (through 'redlining' of parts of the neighbourhood), and lack of schools and community facilities. Low property values followed, alongside continued disinvestment in the housing stock, and subsequently to neighbourhood decline and social problems (Gibson, 2007). Further, the I-5 highway was routed through Albina in 1962, leading to the demolishment of nearly 300 homes along the highway corridor. This was accentuated by later 'urban renewal' projects, such as the Legacy Emanuel Hospital and Moda Center sports arena, which also involved housing demolition and population displacement. The southern neighbourhoods in Albina[1] lost almost two-thirds of their population by 1970, amounting to around 8,000 residents (Cortright, 2021, Davis, 2023). The I-5 was justified, by the state and city authorities, as facilitating the 'regeneration' of the so-called 'blighted' low-income neighbourhood. The highway also provided a connection into the city centre for the higher-income outer suburbanites, usually of White ethnic origin, who had moved away from the inner urban areas as part of the 'white flight' of earlier decades. This was a story replicated in many cities across the USA, with interstate highways and urban renewal justified as important to maintaining the failing tax bases and economies of cities. But, disastrous social equity impacts were overlooked, and city authorities simply reflected the interests of the motor manufacturing and associated development industries and higher-income groups in their comprehensive masterplanning.

Over time, Albina has gentrified, with the displacement of many lower-income, Black residents and others of colour to East Portland, where housing is less expensive, but public transport accessibility much lower. Lower-income residents hence often have lengthy journeys into the city centre or across the metropolitan area. Alongside, there is housing unaffordability in the city centre and the only lower-income residents in central areas are, perversely, the homeless, often with associated drug problems and without healthcare coverage. The 'Portland model' offers sustainable lifestyles in the compact urban centre and inner suburbs, but these are usually only available for higher-income populations, and predominantly of White ethnicity. The higher-quality public transport, walking and cycling networks are mainly used for inner urban trips, rather than for trips in the outer suburbs, which remain car-dependent, with a high level of 'forced' car ownership and usage as there is little other transport provision available.

Incredibly, the lack of consideration of environmental and social public policy goals continues. The current I-5 Rose Quarter Improvement Project is proposed by the Oregon Department of Transportation (ODOT) in the Oregon Transportation Plan (Oregon Department of Transportation, 2023) and Oregon Urban Mobility Strategy (Oregon Department of Transportation, 2022). The project involves increased highway capacity, with an additional highway lane in each direction, between the Fremont Bridge (I-405) and the Banfield Freeway (I-84). The project cost is uncertain, with initial estimates running at $500–750 million, but potentially rising to $1.4–1.9 billion (Cortright, 2023). There is a short section of highway decking through Albina, which offers reduced severance, improved pedestrian and cycle connections, noise reduction and potential use for redevelopment. But, this seems to be offered to placate the local community rather than to improve the environmental and social sustainability of the project. ODOT, it seems, is still being unduly influenced by an obsolete highway-planning agenda, and, presumably, the motor manufacturers and road lobby. ODOT must be aware there are climate change goals and social equity goals, yet still puts forward contradictory increased highway capacity as being important to Portland and Oregon. The promise of electric vehicles (EVs) is unlikely to resolve these problems, as the source power will only be partially renewable, and, of course, EVs lead to congestion and social inequity in similar ways to conventionally fuelled vehicles.

A local community group, The Albina Vision Trust, is supportive of the decking proposals, gaining a $500 million grant through the Federal Reconnecting Communities Fund to act as a community development corporation to produce the masterplan for the area. No More Freeways, a community group, more focused on the environmental sustainability of the transport proposals, contests the highway proposals from ODOT. No More Freeways suggests that increased highway capacity is inconsistent with environmental goals, including climate change, particularly in view of the lack of progress with associated highway tolling plans; and, that the highway plans are inconsistent with the Portland City Plan (City of Portland, 2012). In addition, the traffic modelling is viewed as inaccurate by ignoring induced traffic, hence underestimates projected traffic growth. Indeed, No More Freeways argues that the extent of the planned highway corridor is underplayed and could potentially accommodate much greater highway capacity expansion in future years, with the 160–250 feet (49–76 metres) corridor able to take up to 10 lanes in each direction. No More Freeways continues to fight the proposals, through community workshops, print media and social media releases, and filing lawsuits against ODOT on environmental grounds (No More Freeways, 2024, Wozniacka, 2024).

The discursive formation of the I-5 project, including the knowledge and data used to justify the project, is instructive. ODOT sets the context for the project in the Oregon Transportation Plan (2023, 22), as aiming to 'support all Oregonians by connecting people and goods to places in the most climate-friendly, equitable and safe way'. Hence, there is interactional control and the framing of the problem justifies increased car-based mobility. Environmental and social goals are undefined and impacts are only to be moderated as mobility is enhanced. This language overlooks the inherent contradictions between motorisation and environmental and social equity objectives, hence facilitates increased highway capacity. The Urban Mobility Strategy (2022, 37) states that the I-5 project aims to 'reduce congestion and improve safety' and alleviate the 'biggest traffic bottleneck in Oregon'. The headline indicator uses a congestion metric, suggesting that 'Portland is ranked No.12 in the US for traffic congestion'. The city is therefore positioned as requiring increased highway capacity to reduce congestion. The project is made more acceptable to the local community by adding the small decking element, but this is insignificant relative to the increased traffic volumes that will be facilitated by increased highway capacity. The strategy (5–6) seeks to 'maintain and improve the efficiency of travel of the transportation system, secure funding for multimodal projects that reduce traffic congestion and reduce regional bottlenecks' and states that 'all people should have access to the mode of transportation that works best for them' (in this context, private cars are likely to be the answer). Projects are assessed against goals of 'congestion relief', 'safety', 'climate relief' and 'equity' (15). But, most of the benefits are assumed to result from reduced congestion and stop/start traffic, with the argument even made that this will reduce transport $CO_2$ emissions, overlooking that there are likely to be increased traffic volumes through induced traffic. The equity commentary suggests there is 'robust engagement with underrepresented communities' and that 'project outcomes should reflect community interests' (16). Yet the strategy presents the project options as developed and fixed in design; even with a schedule for implementation. There is some reference to restorative justice (10), as 'the project is committed to leveraging deep ties with the historic Albina community to ensure meaningful involvement with the Black Portlanders'. This is tokenistic, and there is no detail as to how this might be realised and what the impact could be on the proposed project. Albina has gentrified since the 1970s, and much of the Black and wider population of colour has been displaced to East Portland, particularly the lower-income households, as part of the 'suburbanisation of poverty' (Howland, 2020). Understanding how transport projects can more effectively serve such disadvantaged communities needs much greater consideration in city transport strategies. Transport investment and population change, with a constant flux of investment, uneven impacts and change over time is particularly complex to understand. However, projects can be better targeted to particular communities, such as through providing public transport and cycling connections to lower-income neighbourhoods or, indeed, by maintaining affordable housing in central locations. Restorative justice, in terms of substantive social equity (outcomes), would mean reducing vehicular traffic and improving public transport, walking and cycling facilities in Albina and extending these into East Portland, where many of the lower-income groups now live. This would require a suburban and regional rail system due to the distances involved.

The focus of metrics in the Urban Mobility Strategy on issues such as 'travel efficiency', 'reducing congestion' and 'reducing bottlenecks' provides the justification for highway improvement projects. There is strong use of modality (6) in the language of the strategy, for example, expressing that the highway projects are 'an improvement' and that 'congestion is impacting economic competitiveness for the entire state'. This suggests that these are 'realities' or 'truths', yet there is little evidence behind these positions, and readers can easily be misled or manipulated, as they fail to understand that these positions are simply social constructions and political positions, rather than 'facts'. There is poor definition of the strategy objectives or relation to wider environmental, social or even economic public policy goals. There is little consideration of investing in more extensive public transport systems, walking and cycling facilities, including extended public transport services, and reshaping the built environment; or of social equity issues. The envisaged response to lowering transport $CO_2$ emissions is to improve traffic flow and perhaps to charge for use in peak traffic periods, alongside promoting lower emission vehicles. Indeed, there are wider highway projects in the regional strategy, including the I-205 Project, I-5 Boone Bridge and Interstate Bridge Replacement Program. The Oregon Toll Program is proposed to manage traffic levels and help fund the infrastructure projects, but fails to gain political support.

A different strategy could be pursued for the I-5, reflecting the need to prioritise public transport, walking and cycling in the city and region. Much of the I-5 highway infrastructure on the eastern bank is very old and unsightly, creating severance with Southeast Portland, and could be removed. The section through Albina could be decked, but then the I-5 removed from the Moda Center southwards to the I-26. This would open up a new development area on the eastern bank and move through traffic to the I-205 (this could be renamed the I-5 to maintain the integrity of the interstate route, and also be decked through East Portland to remove severance impacts) and I-405. In this way, the priority given to traffic movements would be reversed and public transport, walking and cycling given preference instead. Alongside, there needs to be an effective suburban and regional rail network, including to East Portland, and to surrounding cities. This would be much more consistent with environmental and social public policy goals than highway capacity enhancement. But these types of options are not put forward for discussion by ODOT and many potentially critical strategies and projects remain undiscussed.

## The role of activism

Local environmental and community groups often play a critical role in generating new ideas for transport projects and can lead to changed discourses, or discontinuities, in transport planning. This is often seen with cycling and streetscape reallocation projects, where city authorities can be very slow to invest in high-quality infrastructure. In Portland, and more widely in Oregon, the motorised transport-planning approach is still heavily dominant, despite the large contradictions with environmental and social public policy goals. Local community groups are important

**Figure 23.2** Portland overlooking the Willamette River. The 'Portland model' of urban development, the DISCURSIVE FORMATION, has produced a compact urban centre and attractive city, utilising the industrial heritage, on the banks of the river.

**Figure 23.3** The light rail system, the Metropolitan Area Express (MAX), was built from the 1980s, with five lines; in addition to the Portland Streetcar in the city centre; limited commuter rail and supporting bus services. The MAX routes run from the city centre and university to the inner suburbs and airport, with nearly 100 kilometres of track.

in opposing proposed projects, including highway capacity enhancement projects; indeed, they are likely to be the only way that more progressive approaches are discussed, understood and perhaps eventually adopted. At the national level, in the USA, there is lobbying and sharing of knowledge and experiences through the Freeway Fighters Network (Kimble, 2023). The Congress for New Urbanism (2024) similarly argues for freeway removal projects in different cities, using its series of reports called 'Freeways without Futures'. At the local level, including in Portland, environmental groups such as No More Freeways help to organise debates and grow support for the blocking of highway expansion projects. Local community and environmental activists seek to influence decision-makers at the city level, including city and transport planners and politicians. Sometimes, activists become part of the decision-making process, gaining positions in the city authorities or in consultancies, as city and transport planners, or act as politicians. Hence, there is a gestation and contestation of ideas, with these sometimes becoming more influential as the city authorities adopt the viewpoints offered. The role of the voluntary sector and civil society in contributing to transport planning needs

**Figure 23.4** Cycle network in the city centre – fairly extensive and on-street in the city centre, extending a little into the inner suburbs, including bridge crossings across the river.

to be more thoroughly considered; at the moment, the process is largely reliant on individuals volunteering and spending large amounts of time and effort to be heard. ODOT is continuing to pursue an obsolete highway building agenda in Portland and it is not clear why. The inertia of highway planning, even where there has been a well-known history of freeway removal, means that there is much invested in the motorisation

**Figure 23.5** Waterfront Park. The purpose of government is to influence the actions and behaviours of the population (as the CONDUCT OF CONDUCT). This was positive in terms of the removal of Harbor Drive in the 1970s, as the first significant freeway removal project in the USA, replaced by the Waterfront Park. But the progressive approach to streetspace reallocation has not been maintained.

approach, including funding, technical resources, knowledge and processes. Key governmental, political, business and wider key actors and large cohorts of the public support the highway-planning process, including in influencing the strategies of key agencies such as ODOT. Hence, there are important dimensions of power, which are unclear and untransparent, supporting the current approaches. The governmental bodies are too weak to resist and become 'captured', defending the highway building plans as part of a self-described 'sustainable transport' strategy. Transport planners often fail to adequately problematise the context with which they are faced: in this case, the problematic history of severance and displacement in Albina, and a contemporary requirement to reduce traffic volumes in Portland and the surrounding region on environmental and social grounds. Building any large transport project is not sufficient: project planning and prioritisation needs to fund only the projects that address environmental and social goals.

**Figure 23.6** Redevelopment at the Yards at Union Station and surrounds. Waterfront Park provided space for new urban infill projects.

**Figure 23.7** Highway infrastructure. The legacy of highway building is still evident across the city, with highway infrastructure providing significant severance problems for neighbourhoods. Some of the infrastructure is old and in poor condition, presenting a clear opportunity for removal.

**Figure 23.8** The I-5 interstate highway – the State and City Governments purposively routed the I-5 along the banks of the Willamette River and through the Albina neighbourhood. Built in the 1960s, the highway contributed to housing demolition along the corridor and population displacement. Mobility for higher-income groups was given priority over lower-income homes and neighbourhoods (as the CONDUCT OF CONDUCT).

**Figure 23.9** Historic housing in Albina. Some of the inner suburban neighbourhoods in Portland, such as Hawthorne and Boise, have high-quality housing and connections by public transport and cycling into the city centre. But, still, the neighbourhoods are mostly car-dependent, with ample and free on- and off-street car parking.

**Figure 23.10** Freeway access direct into the city centre and across the metropolitan area. The individual, as the SUBJECT, becomes a vessel for the agenda of motorisation, as travel behaviours and lifestyles continue to be shaped around use of the private car. The very significant adverse impacts of motorisation remain undiscussed by key governmental organisations, as if the environmental and social impacts are unimportant.

**Figure 23.11** Severance problems associated with I-5 interstate highway provision. The REGIME of motorisation is still pervasive, with a very wide corridor of land adversely impacted by the I-5. Yet, the current proposals are to increase highway capacity, leading to increased traffic volumes and $CO_2$ emissions. The environmental and social impacts of motorisation are continually overlooked.

## Note

1. Census tracts 22 and 23 (Eliot and Lloyd), which are the southern neighbourhoods in Albina.

# Part VIII
**Conclusion**

# 24
# Emerging mobility and space?

## Framing sustainable urban mobility

As we have clearly seen, space is produced, rather than simply existing; in other words, it is a product of contemporary social structures and the dimensions of power within these (Lefebvre, 1974). As such, transport can be viewed as a social construct: alongside the technical process of project delivery, transport projects are shaped by and help shape the social context. This includes elements of history, discontinuity, power, truth, ethics and subjectivity. Actors at the institutional level, politicians, businesses, the public, and the cultural norms in a specific context, can all be influential in shaping the transport systems produced. Discourse analysis, using the concepts developed by Foucault (1969, O'Farrell, 2005), helps to uncover and understand how transport systems have been produced and how they are experienced. At the city or project level, or indeed with comparative analysis between cities, the critical factors associated with the discourse of transport can be examined. This has a particular production in each city, but the discursive formation, practice and meaning of transport systems are all impacted by a range of dimensions of discourse. A number of concepts can be examined, such as the episteme, event, practice, problematisation, experience, exclusion, biopower, apparatus, knowledge, techne, normalisation, regime of truth, culture, conduct and subject, etc. Understanding transport provision and travel behaviours through these concepts helps to understand the current state of social practice in transport planning and travel behaviours.

At some stage, heavily motorised transport systems will surely be seen as a time-limited epoch – as widely implemented, but fundamentally flawed – with associated elements of discourse and power. Motorised cities, including where there are limited alternative means of travel and where motorisation is the overriding means of travel, will seek to transition into cities with much more effective public transport, walking and cycling provision. But, as Avila (2014) comments, most cities demonstrate a surprising support for increased mobility, in pursuit of increased materialism and consumption. Economic growth is assumed to be facilitated by increased mobility, yet this largely involves motorisation and economic gains for only the higher-income groups. This overlooks the wider environmental and social impacts of motorisation and the everyday activities and requirements of large cohorts of the population. The poor public transport, walking and cycling networks seen in almost all cities, urban areas and regions, including suburban and exurban areas, reflect the mistakes made, over decades, in transport and city planning. Usually, this resulted from limited prioritisation and funding for public transport, walking and cycling, alongside dispersed built environments – and the private car has filled the gap. This is associated with local discourses on transport planning and travel behaviours, from transport planners, politicians and the public, often supporting the motorised transport systems provided. There are dimensions of knowledge, such as the adopted transport-planning procedures, such as strategy development and project appraisal, which encourage the provision of more motorised mobility. There is often little understanding of the progress required to achieve greater sustainability in travel behaviours and these self-reinforcing mechanisms are difficult to move beyond. Yet, the adverse impacts of motorisation and the benefits of active travel are both so significant that they should be much more widely recognised. As part of this, the extent of the practice of motorisation has to be grasped, including the forms of social practices and power exercised through it. This book seeks to highlight the problems of motorisation, the lack of progress in providing for extensive networks for public transport, walking and cycling, and the contradictions with public policy goals such as climate change and social equity. This occurs in practice, but also in research, where much of the focus is on perceived goals of economic efficiency, informed by concepts of rationality. This can be viewed as a misuse of power and knowledge within transport-planning practice, perpetuating the motorisation agenda.

But there is also hope beyond the poor transport systems found in most cities. We have seen some of the more progressive transport systems developed in wider cities, including the many strengths alongside some weaknesses. Many cities are achieving impressive levels of public transport, walking and cycling. However, the strategies and projects that are considered and implemented in many wider contexts are often strictly bounded – only certain projects and interventions seem possible. The discourses in different cities seem shaped by the understandings and discussions that are held on transport and city planning. This is framed by the historical trajectory that has been followed, strongly influenced by local transport-planning practitioners, politicians and other key actors such as the business community, academics, civil society and the public. The discourses on transport that become accepted influence future possibilities, hence there are elements of 'lock-in' as the discursive framing is continued. Kębłowski and Bassens (2018) have identified broad discourses such as 'neoclassical' perspectives, where the traffic engineering and transport economics-led approaches are used to plan and manage transport systems, including using demand projections and concepts of efficiency. Or, elsewhere, there are 'sustainability' perspectives that encourage new public transport infrastructure, walking and cycling and the compact city concepts of high-density, mixed-use developments around public transport nodes, in other words, versions of the avoid-shift-improve framework (ASIF). Often cities implement a mix of these strategies, but the focus for investment is on specific types of projects and interventions, and others are overlooked. In most cases, the investment in public transport, walking and cycling is not to the degree or quality of intervention that is required to progress towards public policy goals, for example, to significantly reduce transport $CO_2$ emissions. Even in the more progressive cities, strategies usually overlook the distributional impacts and contested nature of project planning and implementation. If we look beyond the (often promotional) narrative that is evident in cities, the vast majority can make major improvements to their transport systems. A fundamental point is the political and cultural system within which transport is developed, often with capitalism and materialism as the overriding and hence structuring factors, giving emphasis to increased consumption, and the facilitation of this. Increased mobility is positioned as a positive feature and close links are assumed between mobility and economic growth. As Lefebvre (1968) discusses, urban space (and transport systems) should not be shaped to facilitate profit maximisation, but instead to improve the lives of citizens. There is a 'right to the city', which is more than the liberty to access urban resources: it is the right to change ourselves by changing the city (Harvey, 2008). Further, the right to the city has been narrowly framed if viewed from contemporary times. Over decades, the neoliberal agenda has been very far-reaching and has shaped the views of many in the population, even to the extent that some argue for retaining the status quo of motorisation dominating everyday life. As Lefebvre (1947) suggests, everyday life, down to the banality of everyday activities (le quotidien), with routines and habits, including travel behaviours, is characterised by political, social and cultural structures. This is framed by the capitalist project, leading to a particular shape of travel and participation in activities. At times, there are also forms of alienation and many people are not well served by the transport systems produced. This reflects that, for many, the impacts of transport systems are negative, the level of participation in activities is not as expected, or the journey experience is poor. We may not think too much of the daily journey in the old, dirty and crowded bus; the Metro journey crammed with other passengers; walking on the uneven or unavailable footpath; cycling amongst the potholes in the busy traffic; or the car journey on the congested highway. Or, indeed, the trip that is not made, as it is deemed too difficult or not worth it. But, all of these experiences reflect the transport systems available, the shape of the built environment, activity participation and the aspirations pursued, and are shaped by the discourses evident on transport in a specific context. There is often a taken-for-granted or resigned nature to the available transport systems, travel behaviours and activity participation. There may be some level of agency at the individual level, but the overriding structuration influences access and experiences, such as the range of possible mobilities relative to time, place and other constraints (Kaufmann et al., 2004). Urban development, infrastructure provision and mobility growth are perceived as key facilitators of materialism and consumption, and this is usually an assumed 'given' for those planning or using transport-planning systems and projects. Yet this epistemological context remains largely undiscussed within transport planning. These ignored dimensions of discourse mean that the critical issues that contextualise and frame policy making and project implementation are overlooked. Ultimately, this leads to poor progress concerning environmental and social equity goals.

Applying a Foucauldian-style discursive framework can help uncover the current status and outcomes of transport provision (Table 24.1) – and, ultimately, to move to more progressive transport systems. Considering discursive features such as formation, practice and meaning, and wider dimensions such as the episteme, experience, exclusion, power, apparatus, knowledge, culture and subject, and so on, can give a much more robust view of the historical and current trajectories and future possibilities. This discursive framework can feasibly be utilised elsewhere – either across or within cities – to help facilitate greater critical understanding of the context under consideration and the likely pathways forward. The premise is to understand the social practices of

**Table 24.1** A discursive framework for understanding transport systems

| Discursive concepts | Application in the selected case study |
|---|---|
| **Key features** | |
| **Discursive formation**: the group of statements, concepts and knowledge available at a given time. | ?? |
| **Discursive practice**: the historically and culturally specific set of rules that define activities and knowledge within a given period or social condition. | ?? |
| **Discursive meaning**: the meaning of the statement depends upon the context within which it emerges. | ?? |
| **Key dimensions** | |
| **History**: the order underlying any given culture at a given time.<br>• Episteme: the unconscious structures underlying the production of scientific knowledge, such as the paradigm.<br>• Event: a component and designation of history. | ?? |
| **Truth**: an event taking place in history, rather than something that exists.<br>• Normalisation: the process of making a truth become evident.<br>• Techne: practical rationality, knowledge and know-how.<br>• Regime of truth: the mechanisms involved in the production of discourses which function as true. | ?? |
| **Power**: a network of relations between actors, rather than a substance.<br>• Governmentality and biopower: the way of administering and managing the population.<br>• Apparatus: institutional, physical and administrative mechanisms or knowledge structures.<br>• Knowledge: historically derived in nature, including use in support of power. | ?? |
| **Discontinuity**: the break and difference in discursive formations, practices and meanings over time.<br>• Practice: the conventional or expected way of doing something.<br>• Problematisation: developing an understanding of the issue that requires a different strategy and solution. | ?? |

| Discursive concepts | Application in the selected case study |
|---|---|
| **Ethics**: the relation of the individual or group to the travel and activities that they may wish to participate in, and to each other.<br>• Experience: the interrelation between normativity, subjectivity and knowledge.<br>• Exclusion: the situation by which societies include or exclude certain groups and individuals. | ?? |
| **Subjectivity**: contestation over the discursive framing and meaning for strategies and projects.<br>• Culture: the organisation of values, accessible to everyone, but also as a mechanism for selection and exclusion.<br>• Conduct: the conduct of conduct as the purpose of government.<br>• Subject: the entity, as self-aware and capable of choosing how to act, yet which often becomes the vessel for the motorisation project. | ?? |

the past and present, including the ideology, and to reveal the past as ideological (Foucault, 1969). In this way, we can consider the foundations for transport-planning practice, and the potential for changing practices and outputs.

This type of interpretative analysis can help transform transport-planning beyond the many current problematic outputs in important ways:

1. Enabling a greater awareness of the current status of transport planning, as defined and shaped by the current historical, political and cultural context, including understanding what 'sustainable transport' might mean in a specific context and how current travel behaviours relate to this.
2. Encouraging a disposition of critique, including dimensions of history, truth, power, discontinuity, ethics and subjectivity, and how these dimensions might relate to transport systems and travel behaviours, in other words, understanding what the current discourses of transport and urban planning mean and how they might change.
3. Leading to a more progressive formulation of strategy and project development for transport and city planning – focused more clearly on the achievement of public policy goals.

## Critique and enlightenment

The requirement, therefore, is to seek to think critically, rather than through prescription. This involves a more effective problematisation of transport systems, travel behaviours and current impacts, including understanding the range of possibilities available for future transport systems and built environments. The lack of problematisation of current transport systems is perceived as being deliberately maintained, by the actors that benefit from the current system, so that the adverse impacts are overlooked (Culver, 2018). Foucault (1978b, 382–3) describes critique as 'a certain manner of thinking, of speaking, likewise of acting, and a certain relation to what exists, to what one knows, to what one does, as well as a relation to society, to culture, to others, and all this one might name the "critical attitude"'. Further, that 'critique only exists in relation with something other than itself: it is an instrument, a means for a future or a truth that it will not know and that it will not be'. With this approach, the type of discourse and truth found in different contexts can be explored, including how practices might be changed. This can involve a questioning of where we are now and what the present reality is that we find. Or, as Foucault (1978b, 39) suggests 'what false idea did knowledge have of itself and to what excessive uses was it exposed, and what domination was it linked to as a consequence?'

Examining the status of transport planning and travel behaviours in such a critical manner will allow an improved understanding of the progress needed to achieve public policy goals in different contexts: 'it is the technique for clearing away things that we take as natural and necessary, in order to begin elsewhere' (Ball, 2019, 2). Critique becomes the means to 'avoiding a fixed and stabilised view of the ever-changing present' (Batters, 2011, 1). It does not declare a 'single truth', but instead helps to analyse discourses of truth, including what is being achieved and what might be achieved. Hence, a clearer understanding is required of the nature of social interactions within society and how these influence transport planning. Transport planners often act apolitically, with an assumption that an effective technical case for new projects will lead to implementation (Hickman and Huaylla Sallo, 2022). But, this assumes that the 'right' projects are being considered, there is an understanding of the possibilities on offer and the progress that needs to be made, and there can be agreement reached over project planning and implementation. The difficulty is that there is often no consensus over the strategy for transport planning; indeed, many of the more attractive possibilities for projects are not even discussed, in many contexts. Interaction between different groups is not always cooperative, and assumptions of shared values and aspirations are misplaced (Simmie, 1974). Varying interpretations are given, such as understanding generally accepted, but fuzzy, concepts such as sustainability and sustainable urban mobility. The same 'objective reality' is interpreted in different ways and, particularly, issues of appropriation are poorly understood, such as who uses the transport system, or not, including how travel is experienced. Markedly different strategies are followed in different contexts, but, ultimately, the great public policy challenges of climate change and social equity are not being achieved in transport at the city and global scales.

The conventional narratives for transport planning in cities differ, but are usually centred on assumptions of continued 'progress', irrespective of what is actually being achieved. The unfortunate realisation is that transport planning has been co-opted into the neoliberal effort (Bahmanteymouri and Mohammadzadeh, 2023). The objectives of capitalism are supported through transport planning, in shaping particular forms of urban development and increased mobility, with projects often conceived for use of the higher-income groups. Competition is used as a central tenet in contracting and operation across many transport domains, and even research, planning and engineering is increasingly carried out by private consultancies, with objectives of profit maximisation, rather than improving the public good.

Very rarely are these assumptions challenged. Almost all cities are not achieving sufficient levels of public transport, walking and cycling to contribute to significantly reduced transport $CO_2$ emissions or improved levels of transport-related social equity – and the lack of progress is continually overlooked. There is rarely evaluation and discussion of progress against policy goals, and there are simplistic assessments made against outdated metrics, such as mobility growth or time savings relative to cost. The empirical study of mobility is usually limited to the analysis of travel behaviours and projects, overlooking the human activity that actually drives the demand for and is involved with travel (Kaufmann, 2011). As a result, transport planning prioritises inappropriate projects.

Sustainable urban mobility has existed as a promising narrative for decades, yet seems very difficult to implement at scale, certainly to any significant level of application. Mostly, sustainable urban mobility remains a marginalised discourse, with a little, well-meaning discussion or promotion, but unsupported by institutions, resources and apparatus, including through information and education. The transport systems, the shape of the built environment and the use of streetspaces in contemporary cities and urban areas are a manifestation of what is regarded as 'normal' in different contexts. This reflects what counts as 'true' for transport and travel behaviours in particular contexts. Transport systems and travel behaviours can be more closely examined, in a more critical manner, across all cities to understand what is required and achievable relative to public policy goals. We can ask the following questions:

- What transport systems are available and why? How do issues such as history, power, truth, ethics and subjectivity affect transport? What does sustainable urban mobility mean in a specific context?
- What do transport planners, engineers, urban planners, politicians and members of the public perceive as the accepted discourse to develop transport systems, and to build cities, in their particular contexts? Are particular projects or interventions for public transport, walking and cycling 'marginally' implemented and contested and why?
- What more progressive transport systems are possible and how might they be achieved? What trajectories are available towards public policy objectives: what projects and packages of projects are required and when? How can progress towards sustainable transport be measured and what mechanisms can ensure effective implementation?

There are many progressive strategies and projects on offer. But, the strong caveat is that we need to be aware of the problems of the current transport systems and the wider possibilities – and to be committed to improving ways of travelling and living. Otherwise, there are many vested interests that will fill the gap, to support and lobby for the dominant motorisation pathway. In the main, these interests have won, over many decades and over multiple contexts. Examining the reality of motorisation shows that this form of travel does not perform well against multiple metrics, but it takes a very careful assessment of energy consumption, $CO_2$ emissions, traffic casualties, the health impacts of inactive travel, and impacts on the built environment, to make this evident. The discursive formation of limited progress towards sustainable urban mobility has become accepted historically and the regimes of truth continually reproduce this truth. But, this position can be viewed as a form of repression on the population, as a mechanism of power. Many of us have waited for a bus that is old and late to turn up, or where a service is withdrawn; paid an extortionate amount for rail travel; attempted to cycle on an uneven road amongst the traffic; or lived in an area with few activities and little public transport provision. All of these are representative of these problems. Some people or institutions will offer narratives about the convenience or even joy of car usage, and that this is what most people choose from transport. But this reflects the fixity of the discursive formation – and it provides no response to the adverse impacts of motorisation.

Foucault (1984) also asks us to consider what is enlightenment, in other words, what are we aiming for? This includes what is thought, said and done, in view of understanding the requirements for change. In terms of transport planning, it is emancipation from inadequate transport systems, and the actors or governance frameworks that produce and reproduce them. The task of rebuilding transport systems and cities will be difficult, as can be seen from the limited progress made over the last decades. There will be much contestation and implementing particular projects may be perceived as 'seemingly impossible' in many contexts. But, I ask you to look at your home city or location and consider whether the transport system could be replanned to better contribute to climate change and social equity social goals. Of course, the answer will be 'yes'. In particular, the nature of power in relation to transport-planning is a very opaque story and difficult to gain evidence on and understand. How the dominative discursive formation of limited progress towards sustainable urban mobility has been so successfully developed over the last century is not so clear, yet we can see that supporting motorisation has been a huge mistake in public policy. Calls for more highways, cheaper fuel, lower taxes, less subsidies for public transport and less state intervention in transport are all frequently heard. The car is presented as the 'obvious' and 'sensible' solution. But, in reality, its success has only been made possible with much initiative and support from a wide variety of actors within the motor vehicle complex (Freund and Martin, 1993, Böhm et al., 2006). The knowledge (connaisance) has been produced to help implement the particular transport systems that were put forward as required. The task of transport planning has historically been to support motorisation, and to gloss over the associated problems. There has been alignment of governmental actors with the objectives of the motor manufacturers, oil producers, the suburban development industry and associated actors, which has helped to maintain their profits. Many city authorities have fallen for this discursive practice and meaning.

But we are now beginning to see that there are many other, more attractive, options on offer – and some progressive examples are discussed in this book. The value of examining case studies in sustainable urban mobility is that they show very different pathways are possible. A critical step is to examine how progressive strategies and projects have been developed, including the elements of discourse associated with them. Positions of discontinuity can then be produced, making transport systems and the resulting travel behaviours more consistent with public policy goals. The task for transport and city planners is to understand the accepted truth in different contexts, but also how the truth is produced, and not produced, including the practices that help shape what is true or false. Hence, the call for empirical research and practice is to put the discourse into transport planning.

As Foucault (1966, 134) suggests, there is a battle for truth and around truth:

> We are studying statements at the limit that separates them from what is not said, in the occurrence that allows them to emerge to the exclusion of others .... The discursive formation is not a developing totality, with its own dynamism or inertia, carrying with it, in an unformulated discourse, what it does not say, what it has not yet said, or what contradicts it at that moment; it is not a rich, difficult germination; it is a distribution of gaps, voids, absences, limits, divisions.

## Employing discourse analysis in transport planning

Transport planning should, therefore, become a broader discipline in terms of the empirical approaches used. The theoretical basis and process of transport planning can be recontextualised, creating different discursive meanings, discursive practices and, eventually, a different episteme for transport planning. Transport planners and wider actors can also become experts in critical analysis, alongside their current strengths in quantitative analysis. This can be part of

> A re-politicisation of everyday (transport-planning) life, by re-opening to question the taken-for-granted and naturalised concepts, practices, relations and social arrangements through which we relate to ourselves and to others. (Ball, 2019)

Table 24.2 summarises the discursive concepts by the case studies discussed previously, including in relation to issues of history, truth, power, discontinuity, ethics and subjectivity. There is much to be learnt from considering the more progressive transport systems, including their strengths and weaknesses – and, also, from the many problems faced elsewhere.

Table 24.2 Dimensions of discourse by case study

| Dimensions of discourse | Example case studies |
|---|---|
| **History**: the order underlying any given culture at a given time. In terms of transport planning, this has been the discursive formations of 'motorisation' and 'modernism' and subsequently of 'marginal progress towards sustainable urban mobility'. These formations were pursued in many cities, with little funding given to public transport, walking and cycling. Key elements include: the episteme and the event. | **Plymouth**: the planning of the city according to the perceived 'progressive' urban- and transport-planning principles of the time meant the city centre was rebuilt in Beaux-Arts style, from the 1940s onwards. The new geometric highway network replaced the old medieval street network. The episteme (the unconscious structure underlying knowledge) was of prioritisation of the car over other modes, including giving the private car direct access to the city centre, high levels of inexpensive car parking and streetspace allocated overwhelmingly to the car. Alongside, transport planning was established as a discipline for producing the framework and facilities for motorisation. Procedures were developed for transport modelling, traffic regulation and control, and extensive highway capacity was produced. The event (a component and designation of history) of modernist planning perpetuates even until today; recent new neighbourhoods, at Sherford and Plymstock Quarry, continue the car-dependent form, despite been labelled by the developers as 'sustainable neighbourhoods'. |

| Dimensions of discourse | Example case studies |
|---|---|
| **Truth**: an event taking place in history, rather than something that exists, in relation to subject, power and knowledge. The discursive formations include 'traffic demand management', 'integrated urban planning and transport' and 'transport and social equity', bringing a revised emphasis to particular strategies. Each illustrate a 'truth' that becomes accepted across wider cities. This truth is often the result of contestation and struggles for power. Key elements include normalisation, the techne and regimes of truth. | **Oxford**: urban highway expansion plans were rejected and a 'balanced' transport strategy was conceived in the 1970s, including traffic restriction and improvements to public transport. This became the techne (practical rationality and knowledge) for many transport strategies. Key measures included city centre pedestrianisation, through traffic restrictions and edge-of-city park and ride sites. | **Freiburg**: environmental-led planning and citizen participation, as the regime of truth (the mechanisms involved in the production of discourse) led to the formation of the strategy incorporating compact city principles, traffic management in the city centre, a refurbished tramway and an extensive cycle network. New urban extensions in Vauban and Rieselfeld, built from 1993–2010, are served by the tramway and cycle network. |
| | **Singapore**: the one-party ruled state gives a particular episteme for policy making, allowing effective strategy implementation, but with limits on participation and contestation. The techne is developed with an integrated 'car-lite' strategy of traffic demand management, public transport and urban planning. Restrictions on vehicle ownership and use were developed from the 1970s, including the certificate of entitlement (COE) and electronic road pricing (ERP). This has limited vehicle numbers to one million across the island, but alongside extensive highway provision. An extensive mass rapid transit (MRT) system and 20 satellite new towns have been developed. | **Bogotá**: The techne of transport planning is shaped to improve accessibility and social equity across the city, including providing projects for the lower income neighbourhoods. The TransMilenio bus rapid transit (BRT) system, built from 2000, has provided a template for BRT systems globally. Crowding problems have led to a Metro system also being planned and built (assumed operational in 2028). Alongside, cycling facilities have been implemented, including the Ciclovía as a weekly event, and the development of a cycle network in the city centre and lower-income neighbourhoods, such as the Alameda El Porvenir route. |

| Dimensions of discourse | Example case studies |
|---|---|
| | **Houten**: the new town was built around an extensive cycle network and traffic restrictions from 1979 – normalising the practice of cycling as the primary means of travel within the town. Houten rail station was built in 1982 and enhanced in 2010, with an additional Castellum station in 2010. The techne of polycentric and compact urban growth, integrated with public transport, is applied through the VINEX regional planning strategy. |
| **Power**: a network of relations between actors, rather than a substance. The discursive practice may include a strong, weak or indifferent governance approach, pursuing strategies that can be productive or repressive. Key elements include governmentality, biopower, apparatus and knowledge. | **Chongqing**: biopower (the management of the population) has been used to produce an extensive public transport network, facilitating a high mode share of trips in the city. An extensive Metro network has been developed in just two decades (since 2004), with 10 lines and 500 kilometres of route length. Two high-speed rail (HSR) stations (Chongqing West and Chongqing East) and another are under construction. The approach to governmentality (administering the population) concerning urban development is of state-led entrepreneurialism, with land sales and city taxes generating funds to contribute to infrastructure investments. |
| | **London, King's Cross**: a new western concourse was built at King's Cross railway station in 2012, following station refurbishment at St Pancras in 2007. This is surrounded by a progressive transit-orientated development (TOD), providing a new neighbourhood with mixed-use, commercial and residential uses. But, there are concerns on the application of the apparatus (institutional mechanisms and knowledge structures) in the development process. This involves the exclusive nature of the development and the lack of affordability in housing provision and commercial space. The private developers dominate the planning process relative to the state. Knowledge (historically derived and in support of power) is used to provide a particular form of urban planning that overlooks social equity issues. |

| Dimensions of discourse | Example case studies |
|---|---|
| | **Rio de Janeiro**: the redevelopment of Porto Maravilha includes the removal of Elevado do Perimetral in 2014 and streetscape improvements at Praça Mauá and Rio Branca in 2015. These are positive developments from some perspectives, but also involve biopower in a specific manner. There are concerns over urban renewal and the distributional impact, as the city is rebranded for the use of higher-income residents and visitors, whilst the everyday experiences of lower-income groups, particularly in the informal neighbourhoods, are not improved. Entrenched social inequity remains difficult to resolve. |
| **Discontinuity**: the break and difference in discursive formations, practices and meanings over time. Key elements include practice and problematisation. | **Utrecht**: the city's transport system has been redesigned from a practice (the expected or conventional way of doing something) of prioritising car usage to a contemporary support for cycling and public transport, including tramway, rail and bus. There has been a gradual removal of the inner-city ring road (built in 1973) and development of an extensive cycle network. Key projects include the Croeslaan cycle route and linear art park; station square cycle parking (with 12,500 cycle parking spaces, opened in 2019); and reinstatement of the city canal at Catharijnesingel. New urban neighbourhoods are developed alongside, such as Leidsche Rijn and Merwede, with good public transport and cycling connections to the city centre. **Copenhagen**: an extensive cycle network has been developed, including segregated cycle routes and also marketing of the city's 'sustainability' credentials. The progressive cycling practice aims to provide cycle journeys that are quick and convenient relative to the car. Cycle projects include: Nørrebrogade and cycle bridges over the harbour, such as Lille Langebro, Cykelslangen and Brygge. The new urban neighbourhood development at Ørestad has excellent public transport and cycling connections, but lacks vitality and street life. |

| Dimensions of discourse | Example case studies |
|---|---|
| | **Malmö:** the deindustrialisation of the city was problematised (the analysis of the way in which an unproblematic set of practices become problematic) and new development planned for the former shipbuilding and industrial docklands at Bo01 and wider Västra Hamnen. A masterplan was implemented from the 2000s onwards. Varied architectural styles were used for housing in the neighbourhoods and the street design prioritised walking and cycling. The masterplanning practice gave emphasis to the use of the spaces in between the buildings, producing a network of spaces to walk, sit, dwell and for social interaction. There was also extensive participation and debate through the 'Creative Dialogue' process. **Dar es Salaam:** the growth of the city in recent decades meant that the informal bus network became chaotic and was unable to cope with mobility demands. The transport issues were problematised and the DART bus rapid transport system implemented, with the first line opening in 2016. The system has gradually been extended, improving the previous unregulated bus system, providing segregated routes, scheduled express and stopping services, and improved waiting areas at bus stations. Use of public transport is possible for journeys across the growing city. |
| | **Delhi:** the Delhi-Meerut regional rapid transit system (RRTS) is the first line of a regional transit system for the National Capital Region and surrounding states. This improves regional connectivity across the metropolitan region, including between Delhi, Ghaziabad and Meerut. RRTS complements the existing Metro system and will be extended into three regional rail lines and potentially more in future years. Travel practices across the city will be dramatically changed with regional movements now possible by public transport, replacing the congested traffic. **Shenzhen:** the city has grown rapidly in recent decades to a population of 17.5 million (in 2023). The transition to a 'low carbon city' has included an extensive Metro system, with 16 lines built in 20 years. The city has also pioneered the practice of electric vehicles (EVs), used for the bus and taxi fleets, and developed as part of the "Ten Cities, Thousands of Vehicles' demonstration programme. Direct funding was provided by the national government, from 2009, and extensive subsidies are used to purchase and run the current fleet of 17,000 electric buses and 22,000 electric taxis. |

| Dimensions of discourse | Example case studies | |
|---|---|---|
| **Ethics**: the relation of the individual or group to the travel and activities that they may wish to participate in, and to each other. This includes the appropriation and use of transport systems. Key concepts include experience and exclusion. | **Manchester**: the development of the Metrolink tramway system, from 1992 onwards, has focused on improving accessibility from the surrounding neighbourhoods and towns into the city centre. Public transport is used to reduce social exclusion (the situation by which societies include or exclude certain groups and individuals). Metrolink has assisted in the regeneration of areas such as Castlefield, Salford Quays, Ancoats and New Islington. The tramway also serves lower-income, outer neighbourhoods, such as Droylsden. Innovative funding mechanisms were required: a Local Transport Fund of £1.5 billion was generated from a council tax levy and wider sources, contributing to a local transport fund. There are some concerns over who uses the transport system (appropriation) and the type of development that is being produced, such as in New Jackson. | **Valenciennes**: a new tramway serves Valenciennes and the surrounding former coal mining areas. The tramway is used to assist in the regeneration of the former industrial areas, improving accessibility and journey experiences (the interrelation between normativity, subjectivity and knowledge) for disadvantaged groups, including those with low incomes and the unemployed. The tram incorporates concessionary travel for the newly employed. Project appraisal is focused on local political goals, including social equity, rather than a more conventional economic efficiency. |
| | **Medellín**: the city sets an innovative example of dealing with very difficult social deprivation problems, extending from the 1980s–90s. The 'social urbanism' approach is shaped with urban development and transport projects, such as the Metro, MetroCable, Metroplús, Encicla and Parques Del Río. The projects are aimed at improving the range of possible mobilities and social equity across different populations groups, including for lower-income groups. Journey experiences are improved, so that activity participation across the city becomes more possible. Citizen commitment is an integral part of project planning and implementation, facilitated by public debates and community participation. Social inequity is difficult to resolve and tensions remain – there are fundamental structural issues that exist. | |

| Dimensions of discourse | Example case studies |
|---|---|
| **Subjectivity**: there may be some contestation over the discursive framing and meaning for strategies and projects. Understanding the levels of subjectivity helps to understand these differences, including the varied viewpoints held, reflecting different experiences and interpretations of reality. Key concepts include culture, conduct and subject. | **London, LTN21**: the production of the low traffic neighbourhood was unsuccessful, and LTN21 was withdrawn in 2021, in the midst of vocal opposition by some residents. The culture (the organisation of values, accessible to everyone, but also as a mechanism for selection and exclusion) of car usage was embedded in some residents' lifestyles and difficult to move beyond. In addition, LTN21 was introduced without effective resources for public participation and project evaluation, or indeed any supporting measures, such as improvements to the cycling network, public transport and public spaces. The challenges of effective participation remain. | **Portland**: the city has a history of public intervention in transport and urban planning, with important environmental and social dimensions. The role of government (in other words, the conduct of conduct, as the purpose of government) is often unclear. For example, there was an early progressive freeway removal project on Harbor Drive in 1974 and the 'Portland model' is well-known in supporting city centre revitalisation, the light rail transit and cycle network. Yet, subsequently, there was support for highway capacity enhancements, including the building of the I-5 highway, institutional racism and associated disinvestment in neighbourhoods such as Albina. Currently, there are Oregon State-led proposals for further capacity enhancements for the I-5 highway. This is being challenged through a freeway fighting campaign, led by environmental and community activists. The subject (the entity, which is seen as self-aware and capable of choosing how to act) becomes, perhaps unwittingly, the vessel for the motorisation project. |

## Types of motility

The case studies can also be considered in terms of the concept of motility (Kaufmann et al., 2004, Shliselberg and Givoni, 2018). There are varied capacities to be mobile and to participate in activities, which act as a factor in social differentiation and integration. Motility includes issues of access (such as the range of possible mobilities according to place, time and other contextual constraints), competence (such as the abilities, knowledge and skills that may influence the level of access) and appropriation (such as how agents, such as individuals, groups, networks and institutions, interpret and act upon perceived or real access and competencies). Each of these contributes to levels of motility, which can range from weak to strong. Mobility is usually not the end goal for individuals, but instead helps to access activities and to engage in social interaction. Motility can cover spatial mobility (the movement of people, goods or information) and social mobility (the transformation in the social position and distribution of resources for individuals or groups). Often motility is assessed at the individual level, but it is also useful to consider at the city level on a comparative basis. Alongside, at the societal level, the transport system is also a factor in progress relative to public policy goals, including environmental, social and well-being objectives.

The case studies include contexts with low levels of motility. These are mostly the car-dependent cities, such as Plymouth, where access by car has improved over decades, including directly into the city centre. But this restricts access by other modes, including public transport (there are only limited bus services and limited segregated lanes); there is some pedestrianisation in the city centre, but cycling facilities are non-existent. Access is therefore improved for car users only, indeed only to a degree as congestion offsets the levels of access in peak periods. Competence issues influence the levels of access, particularly the need to own or have access to a car and a license to drive. Levels of appropriation are low, as many people have constraints on using the car-based system. Hence, overall motility levels are low, as the transport system only works for a limited cohort in the population. If the wider environmental, social and well-being impacts are considered, then the transport system is performing very poorly. We might ask why this situation persists; and why even the recent so-called newer 'sustainable neighbourhoods' are car-dependent? The reasons for this are not so clear, but it seems mostly an issue of inertia, including in knowledge, with transport planners, politicians and the public not able to envisage a different way forward for travel behaviours, hence 'more of the same' is usually the result.

Other cities have medium levels of motility. Singapore provides an example, developing a 'progressive' mobility strategy since the 1970s, with a distinct focus on restricting vehicle ownership, through Electronic Road Pricing (ERP) and the Certificate of Vehicle Entitlement (COE). Alongside, extensive highway capacity and a mass rapid transit (MRT) system have been provided. Hence, there are high levels of access provided through the highway system and MRT system. Urban planning through the new towns ensures residential development is located with good public transport accessibility, usually orientated around an MRT interchange and town centre. But, competence levels are very low concerning access to the highways, as the cost of access is so high, and unaffordable to many, and there are additional requirements such as having a driving license. Competence levels are higher concerning MRT, as fare levels are low, routing and service frequencies are openly available, and almost all can access the public transport system. In terms of overall transport system appropriation, MRT allows all to travel around the city. But, appropriation to the highway system is very inequitable by income; the higher-income groups can use the free-flowing highway network, whilst others use MRT. Hence, in aggregate motility terms, the transport system works relatively well, but only as all can access the extensive public transport system provided. The problematic issue concerns social equity, with only the higher-income group being able to purchase vehicles and use the highway system. There is too much highway space given to private vehicles, and, hence, the aggregate environmental and social performance is not as strong as it could be.

The highest levels of motility are found in cities such as Freiburg, Bogotá and Houten. In Freiburg, there are extensive tram and cycle networks, and a compact urban form, allowing high levels of access for all across the city. Traffic movements are restricted and filtered, so that public transport, walking and cycling are the most convenient modes, including in the city centre, but also from outer neighbourhoods such as Vauban and Rieselfeld to the city centre. There are few competence constraints to accessing these networks, and hence appropriation levels are high. In Bogotá, the TransMilenio, cycle networks and cable cars are purposively designed to serve the lower-income neighbourhoods, giving access between neighbourhoods and to the city centre. Feeder bus services are offered free into the main TransMilenio routes, and the fixed fare means that longer journeys from the outer areas are cheaper on a per-kilometre basis. The Ciclovía provides active mobility across different neighbourhoods in the city, further increasing social interaction and, indeed, health. This is very different practice to most other cities, where infrastructure is targeted towards the areas with perceived greatest economic benefit, such as to and from the wealthy neighbourhoods or city centres. In Bogotá, competency constraints in terms of cost are kept low, hence appropriation is likely to be fairly high, albeit there are wider issues on employment access, such as the relation to education and skills, which might mean that improved access is not necessarily used. Houten provides a different example of high levels of motility. The extensive cycle network, public transport connections and restricted car movements mean that access to the neighbourhood centres, Utrecht and beyond is good; competence levels are high and overall appropriation is high. Motility, therefore, provides a useful concept to apply to transport-planning, allowing an understanding of access and, also, potential usage of the transport systems produced. This takes us beyond economic-led perspectives, which interpret travel as a derived demand and hence lead to projects that increase mobility and reduce travel time. It takes us beyond accessibility-led perspectives, which seek to increase the range of choice of destinations and activities, but fail to consider issues of appropriation and largely still interpret transport as a derived demand. Motility interprets travel (physical and social mobility) within a much wider frame: travel is more than derived and the trip and the activity are linked to personal fulfilment, autonomy and flourishing, leading to happiness and well-being. A mobility index would be useful to develop to help understand these dimensions of sustainable urban mobility. The well-developed public transport systems with inexpensive access and/or the extensive cycling and walking networks provide the higher levels of motility.

An added consideration is the distributional elements of motility. The cities with extensive highway networks are reflective of perpetuating neoliberalism, that is, there are high levels of social inequity in access (and also in terms of environmental impact). Really, it is only the motor manufacturers or higher-income car drivers benefiting from shaping transport systems in this way. Wider social equity problems are evident where public transport networks are limited spatially or in relation to demand, or access costs are high; or, where cycle networks are limited and only serve specific areas, such as city centres, or higher-income neighbourhoods – here, levels of appropriation can remain low. Electrification of vehicle fleets may assist with tailpipe emissions, but less so over the life cycle. Similarly, gentrification and unaffordability resulting from station area redevelopment mean that only specific cohorts are using particular neighbourhoods. There are, therefore, distributional issues to motility that need to be considered, reflecting that certain transport systems may be promoted as contributing to the 'greenification' of cities, but mostly add economic value, overlook social equity impacts, and fail to challenge the dominant and destructive neoliberalism (den Dulk and Buizer, 2024). Motility allows a consideration of transport in terms of not only what we consume (or not) – but also what we become (Shliselberg and Givoni, 2018). Hence, understanding the different elements of motility can become an important part of project planning and implementation. High levels of motility can be an objective for transport planning, as this includes many of the critical components of access, competence and appropriation – all leading from the specific discursive formations and discursive practices evident in each context.

## Positioning discourse analysis within the transport-planning process

Using discourse analysis gives some 'explanatory depth' to understanding transport systems and travel behaviours, therefore helps to understand the current status of transport and travel and to reclaim the accepted discourses in favour of sustainable urban mobility. The challenge is to incorporate forms of qualitative analysis, such as discourse analysis, into the transport-planning process. This can particularly be focused on contexts where there are perceived problems with the current accepted or mainstream discourses, such as the poor progress being made against environmental and social goals, and poor travel options and experiences.

A normative and participatory transport-planning process, including visioning, strategy development and implementation, is suggested (Figure 24.1). The proposed process would replace the current economic-led transport-planning process (Department for Transport, 2024), as applied in contexts such as the UK. This is labelled as objectives-led, but is mostly focused on 'efficiency' objectives. The revised process would give greater consideration of achievement relative to a wider range of public policy goals, and also public acceptability via participation and deliberation. There could be the following stages:

1. Problems and opportunities: analysis to help define the 'problematic', relative to the current status and discursive framing of the transport system and travel behaviours. This can utilise the concepts of discourse analysis as discussed previously, including consideration of history, discontinuity, power, truth, ethics and subjectivity, and related concepts, giving an understanding of the transport system beyond

the conventional focus on travel trends. This can include retrospective archaeological analysis, considering how the transport system, projects and urban developments have been produced, by whom and why?
2. Vision: a projective vision(s) of the future transport system, including key objectives/targets/indicators related to sustainable urban mobility (for a future year, such as 2050, or 25 years forward).
3. Strategy: a plan of action to help implement the vision, using the avoid-shift-improve framework (ASIF) and developing a package of projects and interventions.
4. Appraisal: analysis of the most effective strategy, key projects and programming relative to the vision, key objectives/indicators, and multi-actor views – hence the appraisal involves participatory multi-actor multi-criteria analysis.
5. Implementation programme: the coordinated package of projects, interventions and schedule for implementation, including the likely policy pathways 'backcast' from the agreed vision and objectives to the present day.
6. Evaluation: the final stage of assessment, where progress is assessed relative to the vision and targets, and the strategy is revised, if required, to ensure more effective delivery against the vision.

There is greater attention and analysis given to critically understanding the current status of the transport system and travel behaviours, using qualitative analysis such as discourse analysis. A strategy is developed aiming to achieve policy objectives concerning sustainable urban mobility. The relation to public policy goals becomes clearer as the development of a vision and strategy will be focused on achievement against these objectives. A nested multi-criteria assessment of the strategy ensures there is a robust analysis against sustainability criteria and incorporates multi-actor views (Macharis and Bernardini, 2015, Dean et al., 2019). Projects would only be funded if they contribute to economic, environmental, social and well-being goals. All of these stages can be set within a participatory process of discussion, debate and deliberation, so there is an improved understanding of the rationale for transport planning and greater ownership of the strategies and projects is developed through co-production. The aim of the process is to ensure greater achievement of transport and city planning relative to public policy goals. This takes us beyond the more conventional process of economic-led transport-planning, based on forecasting and efficiency-led project appraisal, prioritising projects mainly in terms of efficiency goals. The extrapolation of

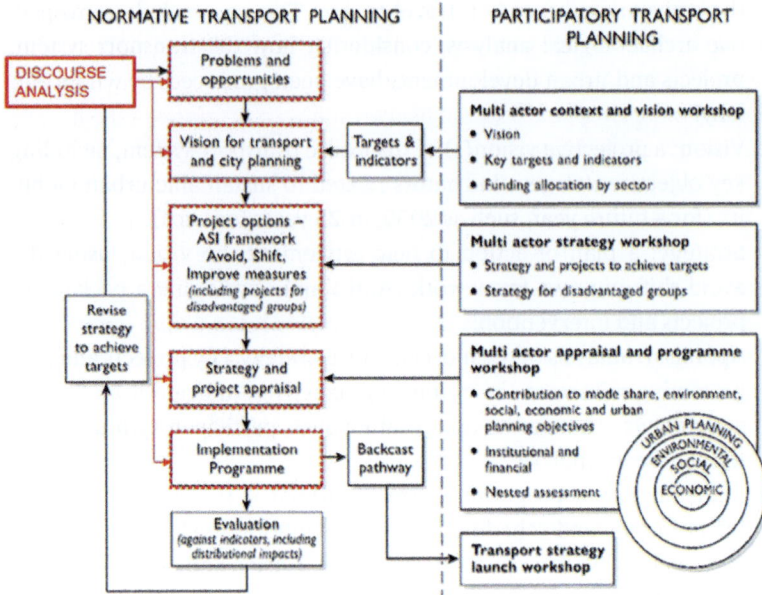

**Figure 24.1** A normative and participatory transport-planning process.

historical trends into a baseline forecast (often used as a reference case), including issues of uncertainty, becomes less important, as the development of the future vision shapes the strategy and programme. Discourse analysis would be most useful in contributing to stage 1 of the process (problems and opportunities), but can also provide contextual information for the later stages 2–5:

## The requirement for participatory and deliberative transport planning

The current rationalist approach to pursuing a planned social order is not achieving contemporary public policy goals. There is contestation over problem identification and the development of strategies and projects. The dimensions of power have become hidden, perhaps on purpose, with transport planning supporting capitalist market objectives and overlooking environmental and social costs. If it is understood that transport-planning is based on conflict and the mediation of different interests, including dimensions of power and uneven resources, then a different set of problems presents itself. Where there is no normative consensus, the technocratic approach breaks down. Social conflict is unlikely to be settled by transport planners acting within the procedures of assumed political neutrality; indeed, the likely result is to extrapolate the historic trends, which are, at times, directly opposed to public policy goals.

Transport planning can become a process achieved through discussion and debate through drawing on communicative and collaborative approaches, such as those found in urban planning (Davidoff, 1965, Habermas, 1984, Healey, 1997). Participants can exchange ideas on what is important and possible, to develop programmes and pathways of action alongside transport-planning officials. There are examples of this type of process in the case studies, perhaps mostly in Freiburg, Malmö and Medellín, where citizen involvement is strong in masterplanning and transport planning. Through these processes, cultures and structures can be transformed – hence the creative dialogue and debate involves the exchange of knowledge and understanding through an open conversation between diverse people. There can be examination of the current discursive formations and potential future options, relative to the policy objectives being tackled. These approaches require the development of more interactive transport-planning procedures, varying according to the type of strategy and projects being developed and the context for implementation. This reflects that forms of knowledge and value do not have objectivity, simply waiting to be discovered, but are actively constituted through social relations. Transport planners, urban planners, politicians and members of the public form interests and values according to their contexts. People have diverse interests and expectations and these are governed through relations of power, including the distribution of resources, infrastructure and wider interventions (Healey, 1997). Public policy and transport planning can reflect this process of mediation, seeking to manage the co-existence of different views, but also to achieve public policy goals. This also means moving away from competitive bargaining towards collaborative policy and project development. This is not a loss of expert knowledge, and reliance on the public to deliver projects, but, instead, a collaborative working process, including views on the current status of transport and the possible futures. This means that transport planning becomes much more than infrastructure development (Nello-Deakin and Nikolaeva, 2021); it is also concerned with elements of history, discontinuity, power, truth, ethics and subjectivity, reflecting an understanding of the socio-political context and possibilities.

This progresses beyond the economic-led transport planning process, as the initial step is to explore why some objectives, rather than others, might be pursued. Discourse analysis becomes useful to problematise the current social status, to offer new trajectories based on discontinuity, and to assess the different viewpoints and positions taken by different actors. The mediation of these competing interests becomes much more central to the procedure of transport planning. And, further, the views of the public(s) become important in the formulation of transport strategy and transport projects, including the technical procedures that need to be thoroughly discussed rather than 'blackboxed'. Participatory and deliberative transport planning can help to provide a space for discussion and debate, fostering an environment for learning about the current condition and the future possibilities, in other words, to help think what it is we do not want to be, and what we might want to be. Hence, the process involves improved awareness of public policy objectives – so the public can choose transport systems and travel behaviours consistent with these. A dialectic materialism approach (Shields, 1999) may be useful in the structure of the discussions, examining opposing views and using questions and answers to clarify ideas and produce solutions, hence raising the level of debate. For example, the use of dialectic involves discussion on the following stages (Maybee, 2020):

- What is the question to be resolved?
- Providing a provisory answer (it seems that ..).
- The principal arguments in favour of the provisory answer.
- Arguments against the provisory answer (perhaps representing the conventional discourse).
- Resolving the question after weighing the evidence.
- Responding to each of the objections.

Transport planning occurs within a particular political context. The discussion on transport systems can recognise the capitalist and market system within which transport operates, so that the overriding structure of society is not ignored. This is a step forward beyond advocacy planning (Davidoff, 1965), where the transport planner acts as the advocate for communities and wider societal values. In this case, the transport planner should help inform, educate and shape societal values, so that the great public policy goals of climate change and social equity are achieved. Transport planning, hence, becomes an interactive process, undertaken in different social contexts, including a technical process of strategy development, project planning and implementation, but also carried out with participation and deliberation. There are debates over the benefits of representative or direct participation, but it is the latter that offers more potential for understanding the different discourses at play.

The conventional approaches to project consultation on transport-planning strategies and projects, such as public consultations, individual/household surveys, workshops, focus groups, written comments and public inquiries, are usually too limited as they do not engage with the issues and people sufficiently (Innes and Booher, 2004). They often do not provide genuine participation as they do not satisfy the people who attend, or do not reach a representative sample of the public. The adversarial manner and/or tokenist nature of the discussions, in relation to actually shaping the projects under consideration, leads to disinterest and lack of engagement amongst the wider public. The mechanisms usually test preconceived options, rather than involving a more thorough consideration of problems and options, leading to co-produced preferred projects and pathways for implementation that are sufficient relative to public policy goals. The principle for effective participation is to demonstrate the transformative power of participation (Forester, 1999), and this should be evident in terms of the strategies or projects developed, or indeed the level of awareness amongst the population. The idea is to move beyond general information-giving and consultation, which occurs in the transport-planning process around project options. Instead, there should be discussion, deliberation, including partnership, delegation of power and citizen control (as famously outlined by Arnstein, 1969), in order to make participatory processes more meaningful. More innovative approaches can be utilised, such as participatory budgeting, public referenda, citizen's juries, citizen's assemblies, multi-stakeholder governance forums, mini-publics, consensus conferences, multi-actor multi-criteria appraisal workshops, deliberative polling and analytic-deliberative approaches (Macharis et al., 2012, Dijk, 2015, Shortall, 2021). The deliberative element is distinct to participation in requiring that participants become more informed on transport planning alongside contributing their views. There is much potential in using electronic methods of debating and polling.

Participatory budgeting is an example of the possibilities for more meaningful participation, where citizens suggest projects and help shape at least a part of the transport funding allocation. Participatory budgeting emerged in Porto Alegre in Brazil in the late 1980s, and has been important in providing infrastructure in poorer parts of the city. More recently, the process been taken up in some European cities, such as in Paris and Madrid, to help prioritise transport projects and budgets. The Madrid process is called 'Decide Madrid', commencing from 2015. This includes an online platform for suggesting and voting on projects. It was introduced following problems of corruption and reduced funding in local government, aiming to improve confidence in local decision-making processes. The Decide Madrid platform has been used in many projects, including the Plaza de España redevelopment, where 27,000 citizens voted on proposals. Other systems exist, such as public referenda in Zürich, which are used to gain public views and support for major project investments. In Germany, there is a pilot project called My Country Talks, run by Zeit Online, an online version of the popular national newspaper. The aim is to encourage dialogue on controversial political issues. People are brought together with different views on a specific topic, to respectfully discuss the issues and to improve the quality of public debate. This can lead to a greater respect for dissenting views, less polarisation in the debate and an improvement in the level of awareness on issues. It has been used on divisive topics, such as immigration, but could potentially also be used to discuss problems in responding to issues such as climate change, social equity and the appropriate policy responses in transport and city planning.

## A more effective transition to sustainable urban mobility?

There can be a shift from understanding transport practice as a simple positivist process of providing infrastructure, which directly leads to travel behaviours. Instead, travel behaviour reflects a wider set of structuring factors. The concept of habitus (Bourdieu, 1972, Maton, 2014) can also be important to transport planning, where the habitus is viewed as the practice of actors (for individuals, institutions and organisations) within the overriding structure of society, in other words, the practice, but including a generative element. The field is the current status of the social arena and an individual's position in the field reflects their specific capital. The resulting practice of travel will be a function of habitus, field and capital. The disposition of people to travel and participate in specific activities reflects the bringing together of structural factors and individual tendency. The concept of choice is viewed as separate to tendency, reflecting that there is often little real 'free choice', with travel behaviours associated with the options and constraints available. For example, the level of access to different elements of the transport system is important to travel, but travel behaviour is also influenced by structural factors such as the type of transport provision (such as extent of network, frequency of services), cost of access, governance systems and funding availability, wider political systems and cultural norms, the shape of the built environment; and, at the individual level, past and present circumstances, attitudes towards transport modes and activity participation, residential location choices, individual political and worldviews and more. Alongside, there are often unclear issues concerning the application and maintenance of power through the process of transport planning. It is these material conditions of existence that shape transport systems and travel behaviours. Hence, transport planning becomes much more complex in theoretical conception – travel behaviours are much more than the function of the attractiveness of origins and destinations.

In addition, transport planning has a wider role than simply being a contributory factor to increasing economic capital (and, further, the poor distribution of economic capital). The transport system and travel behaviours are critical to environmental performance, the societal distribution of activities and resources, well-being and urban planning. There are differential levels of environmental impact, social equity and well-being, all of these occurring by cohort, spatially and over time. The concept of motility (Kaufmann et al., 2004) helps us to understand some of these processes and is similar to the concept of 'disposition' as developed by Bourdieu. We have seen in the case studies discussed earlier that these factors all play out differently in the varied contexts. Understanding these issues of habitus means that transport planning and travel behaviours become much more complex in theory and application than previously understood.

Of course, discourse analysis also has some weaknesses. It is a qualitative approach to understanding societal issues. Hence, there are issues of representation, including the selection of case studies and content for analysis. As with all qualitative analysis, it is suited to some research questions more than others. Discourse analysis cannot be generalised to the wider population, but it does offer a richness to the analysis and can help develop different viewpoints on unjust social practices, including in relation to transport. It can, therefore, generalise in terms of process, in other words helping us to understand that the current status in society might need to change. An example is that the focus on using increased mobility to improve economic efficiency can be revised, as a wider range of contemporary policy objectives have become important, covering economic, environmental, social, well-being and urban planning. Resolving these competing objectives will involve different transport-planning procedures and sets of metrics to judge success and prioritise projects, including revised project appraisal approaches.

Foucault (1969) explains that the history of ideas is relational, that information is not found in a static manner, as a 'pure' state, but instead describes conflicts between the old and new, including repression over those adversely affected and what is and has not been said. In transport planning, contemporary transport systems mostly produce car users (either as car drivers and passengers) with inadequate transport networks for public transport, walking and cycling. Government officials and the wider public are often supportive of the systems produced and the processes that facilitate them. The result, unfortunately, is that there are huge adverse impacts from the resulting motorisation, which are conveniently overlooked. The people who are not included in the transport systems, such as those without access to a car or with little access to public transport (for example, due to poor network coverage or high cost) are also overlooked. This follows from a transport system that is complex in conception: it is the political and societal context, the institutional and administrative mechanisms, the discursive formations and practices that shape transport systems, travel behaviours and their subsequent impacts, including on public policy. There are strong dimensions of power and knowledge, and regimes of truth are developed. The transport systems that are produced lead to particular travel behaviours, activity participation and experiences for people. Attempts to plan and implement different forms of transport systems, without a wider understanding of these wider structuring factors, are likely to fail. But, the contradictions of our current transport systems include the conditions that facilitate the emergence of new discourses. Power can be used in productive ways as well as repressive – and the case studies we have seen, each with their inherent strengths and weaknesses, have shown the possibilities on offer for sustainable urban mobility systems. The purpose of government, including national and city authorities, civil society and wider actors, is to shape the conduct of conduct.

We can ask, what would a city and its region look like without the car as the main means of travel? What might life be like using public transport, walking and cycling for the majority of trips? This gives us an idea of what the transition to sustainable urban mobility might involve. Some of the case studies show glimpses of this future, in different forms. Perhaps most notable are the travel behaviours in Houten or Utrecht, significantly based around the bicycle, walking and public transport; or, indeed, in Bogotá on Sunday mornings during the Ciclovía. The benefits of such active travel and heightened social interactions are obvious to anyone who has the pleasure to experience these cities. We can help develop these new experiences of sustainable urban mobility, and the associated discursive practices, across wider contexts, and this involves changing the political and institutional frameworks that help shape transport systems and travel behaviours. The call is therefore for transport and city planners to progress beyond the current discourses of economic-led motorisation and weak sustainable urban mobility regimes – and the forms of rationality that these rely on and produce. The imperative is to argue for, plan and implement new sustainable urban mobility systems and behaviours. For decades, we have known that we are not achieving sustainable urban mobility, yet continue as if this does not matter. Foucault (1969, 28) tells us that 'the manifest discourse, therefore, is really no more than the repressive presence of what it does not say; and this "not-said" is a hollow that undermines from within all that is said'. Foucault (1988b, 1969) recognises that the status of social practices can be much more flexible than we feel, that people accept as truth, as evidence, the positions that have been established and pursued at particular times in history. However, the evidence can be reconsidered and reshaped. To change something in the minds of people, he suggests, is the role of the intellectual. In relation to transport systems and travel behaviours, this can also be the role of the transport and city planner: we can seek to understand the transport and city planning practices of the past, present and future, including their ideology, and to reveal the trajectories as ideological and changeable.

# References

Abercrombie, P. and Paton Watson, J. 1943. *A Plan for Plymouth*. Plymouth: Plymouth City Council.
Ackerman, F. and Heinzerling, L. 2004. *Priceless: On knowing the price of everything and the value of nothing*. New York: New Press.
Afonin, A. 2021. *Discourses on Low Traffic Neighbourhoods: Understanding the impacts of implementing an LTN in Ealing, London* (unpublished MSc dissertation). London: Bartlett School of Planning, UCL.
African Development Bank. 2015. *Dar es Salaam Bus Rapid Transit: Environmental and Social Impact Assessment Summary*. Dar es Salaam: ADB.
Arnstein, S. 1969. 'A ladder of citizen participation'. *Journal of the American Institute of Planners*, 35: 216–24.
Asian Development Bank. 2017. *Guidelines for the Economic Impacts of Projects*. Manila: ADB.
Asian Development Bank. 2020. *Delhi–Meerut Regional Rapid Transit System Investment Project: Economic analysis*. Manila: ADB.
Austin, G. 2013. 'Case study and sustainability assessment of Bo01, Malmö, Sweden'. *Journal of Green Building*, 8: 34–50.
Avila, E. 2014. *The Folklore of the Freeway: Race and revolt in the modernist city*. Minneapolis: University of Minnesota Press.
Bahmanteymouri, E. and Mohammadzadeh, M. 2023. '"Neoliberalism is dead": Traversing neoliberal planning education is an exigency'. *Policy Futures in Education*, 22: 625–41.
Ball, S. 2019. 'A horizon of freedom: Using Foucault to think differently about education and learning'. *Power and Education*, 11: 132–44.
Banister, D. 2002. *Transport Planning*. London: Spon.
Bao, H. X. H., Li, L. and Lizieri, C. 2019. 'City profile: Chongqing (1997–2017)'. *Cities*, 94: 161–71.
Barter, P. 2013. 'Singapore's mobility model: Time for an update?' In *Institute for Mobility Research* (ed.) *Megacity Mobility Culture: How cities move on in a diverse world*. Berlin: Springer.
Barter, P. 2019a. 'Singapore'. In Pojani, D., Corcoran, J., Sipe, N., Mateo-Babiano, I. and Stead, D. (eds) *Parking: An international perspective*. Amsterdam: Elsevier.
Barter, P. 2019b. 'Singapore's changing relationship with cars'. In Hamnett, S. and Yuen, B. (eds) *Planning Singapore: The experimental city*. London: Routledge.
Batters, S. 2011. 'Care of the self and the will to freedom: Michel Foucault, critique and ethics'. Accessed January 2024. www.semanticscholar.org/author/Stephanie-M.-Batters/115612359.
Berger, P. and Luckman, T. 1966. *The Social Construction of Reality: A treatise in the sociology of knowledge*. New York: Anchor Books.
Bicycle Dutch. 2011. 'How the Dutch got their cycleways'. Accessed August 2022. https://bicycledutch.wordpress.com/2011/10/20/how-the-dutch-got-their-cycling-infrastructure/.
Bicycle Dutch. 2016. 'Motorway removed to bring back the original water'. Accessed December 2019. https://bicycledutch.wordpress.com/2016/01/05/motorway-removed-to-bring-back-original-water/.
Bicycle Dutch. 2019a. 'Finally fully open: Utrecht's huge bicycle parking garage'. Accessed December 2019. https://bicycledutch.wordpress.com/2019/08/20/finally-fully-open-utrechts-huge-bicycle-parking-garage/.
Bicycle Dutch. 2019b. 'From 4-lane road to linear art park'. Accessed December 2019. https://2bicycledutch.wordpress.com/2019/04/17/croeselaan/.
Bishop, P. and Williams, L. 2016. *Planning, Politics and City Making: A case study of King's Cross*. London: RIBA Publishing.

Böhm, S., Jones, C., Land, C. and Paterson, M. 2006. 'Part One, Conceptualizing Automobility: Introduction: Impossibilities of automobility'. *The Sociological Review*, 54: 1–16.

Boughton, J. 2013. 'Municipal Dreams: A plan for Plymouth: Our first great welfare-state city'. Accessed June 2025. https://municipaldreams.wordpress.com/.

Bourdieu, P. 1972. *Outline of a Theory of Practice*. Cambridge: Cambridge University Press.

Bourdieu, P. 1991. *Language and Symbolic Power*. Cambridge: Polity Press.

Brand, P. and Dávila, J. 2011. 'Mobility innovation at the urban margins'. *City*, 15: 647–61.

Brown, G. and Yule, G. 1983. *Discourse Analysis*. Cambridge: Cambridge University Press.

Buczynski, A. 2018. *Filtered Permeability on Cycle Highway C95 in Copenhagen*. Brussels: European Cyclists' Federation.

Buehler, R. and Pucher, J. 2011. 'Sustainable transport in Freiburg: Lessons from Germany's environmental capital'. *International Journal of Sustainable Transportation*, 5: 43–70.

Bus Rapid Transit Centre of Excellence. 2023. Global BRT data. Santiago. Accessed December 2023. http://brtdata.org.

Cairns, S., Atkins, S. and Goodwin, P. 2002. 'Disappearing traffic: The story so far'. *Municipal Engineer*, 15: 13–22.

CEIC Data. 2024. 'India Number of Registered Motor Vehicles: NCT of Delhi'. Accessed February 2024. www.ceicdata.com/en/india/number-of-registered-motor-vehicles-nct-of-delhi.

Chalkley, B. and Goodridge, J. 1991. 'The 1943 Plan for Plymouth: War-time vision and post-war realities'. In *Plymouth: Maritime city in transition*, edited by B. Chalkley, D. Dunkerley and P. Gripaios. London: David & Charles.

Chen, C.-L. and Hickman, R. 2020. 'Impacts of high-speed rail: Hubs, linkages, and development'. *Built Environment*, 46: 337–41.

Chen, C.-L., Hickman, R. and Saxena, S. 2014. *Improving Interchanges: Towards better multimodal hubs in the PRC*. Manila: Asian Development Bank.

Chongqing Municipal Government. 2020. *The 14th Five-Year Plan of Chongqing's Urban Infrastructure Construction (2021–2025)*. Chongqing: CMG.

Church, A., Frost, M. and Sullivan, K. 2000. 'Transport and social exclusion in London'. *Transport Policy*, 7: 195–205.

City of Copenhagen. 2011. *Good, Better, Best: The city of Copenhagen's bicycle strategy, 2011–2025*. Copenhagen: City of Copenhagen, Traffic Department.

City of Copenhagen. 2018. *The Bicycle Account 2018*. Copenhagen: CoP.

City of Portland. 2012. *The Portland Plan*. Portland: CoP.

City of Portland and Multnomah County. 2015. *Climate Action Plan*. Portland: CoP.

Colville-Anderson, M. 2018. *Copenhagenize: The definitive guide to global bicycle urbanism*. Washington DC: Island Press.

Congress for the New Urbanism. 2009. 'Portland's harbor drive'. Accessed March 2024. https://web.archive.org/web/20130420154442/www.cnu.org/highways/portland.

Congress for the New Urbanism. 2024. 'Freeways without futures'. Washington DC: CNU. Accessed April 2024. www.cnu.org/our-projects/highways-boulevards/freeways-without-futures.

Cortright, J. 2021. 'How ODOT destroyed Albina: The I-5 Meat Axe'. City Observatory. Accessed May 2024. https://cityobservatory.org/how_odot_destroyed_albina_par2/.

Cortright, J. 2023. 'Another exploding whale: ODOT's freeway widening cost quadruples'. City Observatory. Accessed May 2024. https://cityobservatory.org/rosequarter_cost_overrun23/.

Crippa, M., Guizzardi, D., Banja, M., Solazzo, E., Muntean, M., Schaaf, E., Pagani, F., Monforti-Ferrario, F., Olivier, J., Quadrelli, R., Risquez Martin, A., Taghavit-Moharamli, P., Grassi, G., Rossi, S., Oom, D., Branco, A., San-Miguel, J. and Vignati, E. 2022. $CO_2$ *Emissions of All World Countries*. Ispra: JRC/IEA/PBL.

Crotty, M. 1998. *Foundations of Social Research: Meaning and perspective in the research process*. London: SAGE Publications.

CROW. 2016. *Design Manual for Bicycle Traffic*. Utrecht: CROW.

Culver, G. 2017. 'Mobility and the making of the neoliberal "creative city": The streetcar as a creative city project?'. *Journal of Transport Geography*, 58: 22–30.

Culver, G. 2018. 'Death and the car: On (auto)mobility, violence, and injustice'. *ACME: An International Journal for Critical Geographies*, 17: 144–70.

Dalkmann, H. and Brannigan, C. 2007. *Transport and Climate Change, Sourcebook Module 5e*. Bonn: Deutsche Gesellschaft für Internationale Zusammenarbeit (GIZ) GmbH.

David, R. 2023. 'Albina's destruction. Publishing prejudice'. *The Oregonian*. Accessed May 2024. https://projects.oregonlive.com/publishing-prejudice/whitewashing-destruction.

Davidoff, P. 1965. 'Advocacy and pluralism in planning'. *Journal of the American Institute of Planners*, 31: 331–8.
Dávila, J. D. (ed.). 2013. *Urban Mobility and Poverty: Lessons from Medellín and Soacha, Colombia*. London; Medellín: DPU, Universidad Nacional de Colombia.
Dean, M., Hickman, R. and Chen, C.-L. 2019. 'Testing the application of participatory MCA: the case of the South Fylde Line'. *Transport Policy*, 73: 62–70.
Deer, C. 2012. 'Doxa'. In *Pierre Bourdieu: Key Concepts*, edited by M. Grenfell. Abingdon: Routledge.
Den Dulk, L. S. and Buizer, M. 2024. 'The shadow of urban greening initiatives: A pluralistic discursive space approach to the High Line and the BeltLine'. *Geoforum* 149: 103938.
Department for the Environment and Department for Transport. 1992. 'Residential roads and footpaths: Layout considerations'. *Design Bulletin*, 32. London: HMSO.
Department for Transport. 2023. *The Plan for Drivers*. London: DfT.
Department for Transport. 2024. 'Website for Transport Analysis Guidance (WebTAG)'. Accessed May 2024. www.gov.uk/guidance/transport-analysis-guidance-tag.
Deutsche Gesellschaft für Internationale Zusammenarbeit (GIZ). 2018. 'Trends and challenges in electric bus development in China'. *Sustainable Transportation in China Blog*, April. Accessed January 2023. https://web.archive.org/web/20180907001358/http://www.sustainabletransport.org/archives/5770.
Diao, M. 2018. 'Towards sustainable urban transport in Singapore: Policy instruments and mobility trends'. *Transport Policy*, 81: 320–30.
Dijk, M. 2015. 'From government to multi-stakeholder governance for sustainable mobility'. In *Transitions to Sustainability*, edited by F. Mancebo and I. Sachs. Dordrecht: Springer Netherlands.
Dryzek, J. 1997. *The Politics of the Earth: Environmental Discourses*. Oxford: Oxford University Press.
Dudley, G. 2013. 'Why do ideas succeed and fail over time? The role of narratives in policy windows and the case of the London congestion charge'. *Journal of European Public Policy*, 20: 1139–56.
Ealing Council. 2021. *FAQs – Low Traffic Neighbourhoods (LTN) to Support Social Distancing*. Ealing: Ealing Council.
Eckhouse, B. and Dlouhy, J. 2021. 'Electric buses are poised to get a U.S. infrastructure boost'. Accessed June 2025. www.bloomberg.com/news/newsletters/2021-08-13/electric-buses-are-poised-to-get-a-u-s-infrastructure-boost.
Edwards, M. 2010. 'King's Cross: Renaissance for whom?' In *Urban Design and the British Urban Renaissance*, edited by J. Punter. Abingdon: Routledge.
Elliott, J. and Earwaker, R. 2021. *Renters on Low Incomes Face a Policy Black Hole*. York: Joseph Rowntree Foundation.
Engels, F. 1845. *The Condition of the Working-Class in England in 1844*. Leipzig: Otto Wigand (translated into English 1886).
English Heritage. 2002. *Manchester: The warehouse legacy*. London: English Heritage.
Fairclough, I. and Fairclough, N. 2012. *Political Discourse Analysis*. Abingdon: Routledge.
Fairclough, N. 1992a. *Discourse and Social Change*. Cambridge: Polity Press.
Fairclough, N. 1992b. Introduction. In *Critical Language Awareness*, edited by N Fairclough. London: Longman.
Fairclough, N. 2010. *Critical Discourse Analysis: The critical study of language*. 2nd edition. Abingdon: Routledge.
Fairclough, N., Graham, P., Lemke, J. and Wodak, R. 2004. 'Introduction'. *Critical Discourse Studies*, 1: 1–7.
Faubion, J. (ed.). 1994. *Power: Essential works of Foucault, 1954–1984*. Volume 3. London: Editions Gallimard (Penguin Books, 2002).
Field, B. 1992. 'Singapore's new town prototype: A textbook prescription?' *Habitat International*, 16: 89–101.
Finn, P. 2022. 'Low-traffic neighbourhoods in Ealing: contested policy making in a polycentric governance environment'. Accessed June 2025. https://doi.org/10.31124/advance.20120411.v1.
Flyvbjerg, B. 1998. *Rationality and Power: Democracy in practice*. Chicago; London: University of Chicago Press.
Flyvbjerg, B. 2011. 'Case study'. In *The Sage Handbook of Qualitative Research*, edited by N. Denzin and S. Lincoln. Thousand Oaks: Sage.
Forester, J. 1999. *The Deliberative Practitioner: Encouraging Participatory Planning Processes*. Cambridge, MA: MIT Press.
Foucault, M. 1961. *Madness and Civilization: A history of insanity in the age of reason*. London: Tavistock.
Foucault, M. 1966. *The Order of Things: An archaeology of the human sciences*. London: Tavistock.

Foucault, M. 1967. 'Nietzsche, Freud, Marx'. In *Aesthetics, Method and Epistemology: The essential works of Michel Foucault 1954–1984*, edited by J. Faubion. Harmondsworth: Allen Lane (1998).
Foucault, M. 1969. *The Archaeology of Knowledge*. London: Tavistock.
Foucault, M. 1975. *Discipline and Punish: The birth of the prison*. London: Allen Lane.
Foucault, M. 1977. 'Truth and power'. In *Power/Knowledge: Selected interviews and other writings 1972–1977*, edited by C. Gordon. New York: Vintage Books.
Foucault, M. 1978a. *The History of Sexuality. Volume I: An Introduction*. New York: Pantheon Books.
Foucault, M. 1978b. 'What is critique?' In *What is Critique? And the Culture of the Self*, edited by H.-P. Fruchaud, D. Lorenzini, A. Davidson and C. O'Farrell. Chicago: University of Chicago Press (2024).
Foucault, M. 1980. 'Interview with Michel Foucault'. In *Michel Foucault. Power: Essential works of Foucault, 1954–1984*, edited by J. Faubion. Volume 3. London: Editions Gallimard, 1994 (Penguin Books, 2002).
Foucault, M. 1982a. 'Space, knowledge and power'. In *Michael Foucault. Power: Essential works of Foucault, 1954–1984*, edited by J. Faubion. London: Penguin Books.
Foucault, M. 1982b. 'The subject and power'. In *Michel Foucault: Power: Essential works of Foucault, 1954–1984*, edited by J. Faubion. London: Penguin Books.
Foucault, M. 1983a. 'The culture of the self'. In *What is Critique? And the Culture of the Self*, edited by H.-P. Fruchaud, D. Lorenzini, A. Davidson and C. O'Farrell. Chicago: University of Chicago Press (2024).
Foucault, M. 1983b. *Fearless Speech*, edited by J. Pearson. Los Angeles: Semiotext.
Foucault, M. 1984. 'What is enlightenment?' In *The Foucault Reader*, edited by P. Rabinow. London: Penguin (1991).
Foucault, M. 1988a. 'Technologies of the self'. In *Technologies of the Self: A seminar with Michel Foucault*, edited by L. Martin, H. Gutman and P. Hutton. London: Tavistock.
Foucault, M. 1988b. 'Truth, power, self: An interview with Michel Foucault'. In *Technologies of the Self: A seminar with Michel Foucault*, edited by L. Martin, H. Gutman and P. Hutton. London: Tavistock.
Foucault, M. 1991. 'Governmentality'. In *The Foucault Effect: Studies in governmentality*, edited by G. Burchell, C. Gordon and P. Miller. Chicago: Chicago University Press.
Foucault, M. 1994. *Dits et Écrits IV*. Paris: Gallimard.
Foucault, M. 1997. *Society Must Be Defended: Lectures at the Collège de France, 1975–1976*. New York: St. Martin's Press.
Foucault, M. 2001. *The Hermeneutics of the Subject: Lectures at the Collège de France, 1981–1982*. Paris: Gallimard Seuil.
Foucault, M. 2008. *The Birth of Biopolitics: Lectures at the Collège de France, 1978–1979*, translated by G. Burchell. New York: St Martin's Press.
Freund, P. and Martin, G. 1993. *The Ecology of the Automobile*. Montreal: Black Rose Books.
Galle, M. and Modderman, E. 1997. 'VINEX: National Spatial Planning Policy in the Netherlands during the Nineties'. *Netherlands Journal of Housing and the Built Environment*, 12: 9–35.
Gehl Architects. nd. *Bo01 and Western Harbour Case Study*. Copenhagen: Gehl Architects.
Gehl, J. 1987. *Life Between Buildings: Using public space*. New York; Wokingham: Van Nostrand Reinhold.
Gehl, J. 2010. *Cities for People*. Washington DC: Island Press.
Gibson, K. 2007. 'Bleeding Albina: a history of community disinvestment, 1940–2000'. *Transforming Anthropology*, 15: 3–25.
Giddings, B., Hopwood, B. and O'Brien, G. 2002. 'Environment, economy and society: fitting them together into sustainable development'. *Sustainable Development*, 10: 187–96.
Gilbert, A. 2008. 'Bus rapid transit: Is Transmilenio a miracle cure?' *Transport Reviews*, 28: 439–67.
Glass, R. 1964. 'Introduction'. In *Centre for Urban Spaces* (ed.) London: Aspects of Change. London: MacGibbon and Kee.
Godfrey, J. 2021. 'A road through the meadow'. Christ Church blog. Oxford. Map source: digital. Accessed November 2021. bodleian.ox.ac.uk.
Goodling, E., Green, J. and McClintock, N. 2015. 'Uneven development of the sustainable city: Shifting capital in Portland, Oregon'. *Urban Geography*, 36: 504–27.
Goodwin, P., Hallett, S., Kenny, F. and Stokes, G. 1991. *Transport: The new realism: Report to Rees Jeffreys Road Fund*. Oxford: University of Oxford, School of Geography and Environment.
Gordon, C. (ed.) 1980. *Michael Foucault. Power/Knowledge: Selected interviews and other writings, 1972–1977*. New York: Vintage Books.

Gössling, S. 2013. 'Urban transport transitions: Copenhagen, city of cyclists'. *Journal of Transport Geography*, 33: 196–206.

Gössling, S., Schweiggart, N., Nieuwenhuijsen, M., McEachan, R. R. C. and Khreis, H. 2024. 'Urban transport system changes in the UK: In danger of populism?' *Cities*, 153: 105273.

Gould, J. 2010. *Plymouth: Vision of a modern city*. Swindon: English Heritage.

Greater London Authority. 2004. *London Plan: Spatial development strategy for London*. London: GLA.

Guiver, J. 2007. 'Modal talk: Discourse analysis of how people talk about bus and car travel'. *Transportation Research Part A*, 41: 233–48.

Guzman, L., Oviedo, D. and Cardona, R. 2018. 'Accessibility changes: Analysis of the integrated public transport system of Bogotá'. *Sustainability*, 10: 3958.

Gwilliam, K. 2002. *Cities on the Move: A world bank urban transport strategy review*. Washington DC: World Bank.

Habermas, J. 1984. *The Theory of Communicative Action: Reason and the rationalization of society*. Cambridge: Polity Press.

Hackworth, J. and Smith, N. 2001. 'The changing state of gentrification'. *Tijdschrift voor Economische en Sociale Geografie*, 92: 464–77.

Hajer, M. and Versteeg, W. 2005. 'A decade of discourse analysis of environmental politics: Achievements, challenges, perspectives'. *Journal of Environmental Policy & Planning*, 7: 175–84.

Hall, P. 2014. *Good Cities, Better Lives*. Abingdon: Routledge.

Hamiduddin, I. 2015. 'The car in the neighbourhood: Residential design and social outcomes in southern Germany'. In *Handbook on Transport and Development*, edited by R. Hickman, M. Givoni, D. Bonilla, and D. Banister. Cheltenham: Edward Elgar.

Hamilton-Baillie, B. 2008. 'Shared space: Reconciling people, places and traffic'. *Built Environment*, 34: 161–81.

Hardy, C. 2014. 'Hysteresis'. In *Pierre Bourdieu: Key concepts*, edited by M. Grenfell. Abingdon: Routledge.

Harvey, D. 1973. *Social Justice and the City*. Baltimore: Johns Hopkins University Press.

Harvey, D. 2005. *A Brief History of Neo-Liberalism*. Oxford: Oxford University Press.

Harvey, D. 2008. 'The right to the city'. *New Left Review*, 53.

Hay, A. 1995. 'Concepts of equity, fairness and justice in geographical studies'. *Transactions of the Institute of British Geographers*, 20: 500–8.

Healey, P. 1997. *Collaborative Planning: Shaping places in fragmented societies*. Basingstoke: Macmillan.

Hickman, R. 2017. 'Sustainable travel or sustaining growth?' In *The Routledge Handbook of Transport Economics*, edited by J. Cowie and S. Ison. Abingdon: Routledge.

Hickman, R. 2019. 'The gentle tyranny of CBA in transport appraisal'. In *Transport Matters*, edited by I. Docherty and J. Shaw. Bristol: Policy Press.

Hickman, R. 2021. 'LTNs and Lefebvre'. *Town & Country Planning*, November: 365–7.

Hickman, R. 2023. 'Car advertising and environmental greenwashing'. In *Discourse Analysis in Transport and Urban Development: Interpretation, controversy and diversity*, edited by R. Hickman and C. Hannigan. Cheltenham: Edward Elgar.

Hickman, R. & Afonin, A. 2022. 'Transport $CO_2$ mitigation and the production of low traffic neighbourhoods: Lessons from London'. In *Climate Change Mitigation: Policies and Lessons for Asia*, edited by D. Azhgaliyeva and D. Rahut. Tokyo: ADBI.

Hickman, R. and Afonin, A. 2025. 'Understanding the opposition to low traffic neighbourhoods'. In *Handbook of Transportation and Public Policy*, edited by A. Perl, R. Ray and L. Reardon. Cheltenham: Edward Elgar.

Hickman, R. and Banister, D. 2014. *Transport, Climate Change and the City*. Abingdon: Routledge.

Hickman, R. and Banister, D. 2019. 'Transport and the environment'. In *A Research Agenda for Transport Policy*, edited by J. Stanley and D. Hensher. Cheltenham: Edward Elgar.

Hickman, R., Cao, M., Mella Lira, B., Fillone, A., Biona, J. and Lopez, N. 2017a. 'Understanding capabilities, functionings and travel in high and low income neighbourhoods in Manila'. *Social Inclusion*, 5: 161–74.

Hickman, R. and Dean, M. 2018. 'Incomplete cost – incomplete benefit analysis in transport appraisal'. *Transport Reviews*, 38: 689–709.

Hickman, R., Fremer, P., Breithaupt, M. and Saxena, S. 2011. *Changing Course in Sustainable Urban Transport: An illustrated guide*. Manila: Asian Development Bank.

Hickman, R., Garcia, M. M., Arnd, M. and Peixoto, L. F. G. 2021a. 'Euston station redevelopment: Regeneration or gentrification?'. *Journal of Transport Geography*, 90: 102923.
Hickman, R. and Hannigan, C. (eds). 2023. *Discourse Analysis in Transport and Urban Development: Interpretation, controversy and diversity*. Cheltenham: Edward Elgar.
Hickman, R. and Huaylla Sallo, K. 2022. 'The political economy of streetspace reallocation projects: Aldgate Square and Bank Junction, London'. *Journal of Urban Design*, 27: 397–420.
Hickman, R., Lu, P. and Botermans, A. 2025. 'The discourse of cycling in Houten'. *Journal of Urban Design*: 1–21.
Hickman, R., Mella Lira, B., Givoni, M. and Geurs, K. (eds). 2019. *A Companion to Transport, Space and Equity*. Cheltenham: Edward Elgar.
Hickman, R., Moubray, K. and Hannigan, C. 2021b. '"Make her yours": The shape and use of car advertising'. London: Bartlett School of Planning, UCL.
Hickman, R. and Osborne, C. 2017. *Sintropher Executive Summary, Interreg IVB*. London: UCL.
Hickman, R., Smith, D., Moser, D., Schaufler, C. and Vecia, G. 2017b. *Why the Automobile Has No Future: A Global Impact Analysis*. Hamburg: Greenpeace Germany.
Hidalgo, D. and Carrigan, A. 2010. 'BRT in Latin America high capacity and performance, rapid implementation and low cost'. *Built Environment*, 36: 283–97.
Hidalgo, D., Pereira, L., Estupiñán, N. and Jiménez, P. L. 2013. 'TransMilenio BRT system in Bogota, high performance and positive impact – Main results of an ex-post evaluation'. *Research in Transportation Economics*, 39: 133–8.
Howland, S. 2020. "I should have moved somewhere else": The impacts of gentrification on transportation and social support for black working-poor families in Portland, Oregon' (unpublished PhD dissertation). Portland State University.
Innes, J. and Booher, D. 2004. 'Reframing public participation: strategies for the 21st century'. *Planning Theory & Practice*, 5: 419–36.
INRIX. 2021. *INRIX Global Traffic Scorecard*. Kirkland: INRIX.
Institute for Transportation and Development Policy (ITDP). 2010. *Case Study: Houten*. Utrecht: ITDP Europe.
International Energy Agency. 2015. *World Energy Outlook*. Paris: IEA.
International Energy Agency. 2016. *Energy Technology Perspectives*. Paris: IEA.
International Organization of Motor Vehicle Manufacturers (OICA). 2020. Vehicle Usage Statistics. Accessed March 2025. www.oica.net/category/vehicles-in-use/.
Jacobs, A. 1993. *Great Streets*. Cambridge, MA: MIT Press.
Jacobs, J. 1961. *The Death and Life of Great American Cities*. New York: Random House.
Jain, J. and Lyons, G. 2007. 'The gift of travel time'. *Journal of Transport Geography*, 16: 81–9.
Jones, P. 1989. 'Oxford: An evolving transport policy'. *Built Environment*, 15: 231–43.
Kaufmann, V. 2002. *Re-thinking Mobility: Contemporary Sociology*. Aldershot: Ashgate.
Kaufmann, V. 2011. *Rethinking the City: Urban dynamics and motility*. Lausanne: EPFL Press: Routledge.
Kaufmann, V., Bergman, M. and Joye, D. 2004. 'Motility: Mobility as capital'. *International Journal of Urban and Regional Research*, 28: 745–56.
Kaufmann, V., Dubois, Y. and Ravalet, E. 2018. 'Measuring and typifying mobility using motility'. *Applied Mobilities*, 3: 198–213.
Kębłowski, W. and Bassens, D. 2018. '"All transport problems are essentially mathematical": The uneven resonance of academic transport and mobility knowledge in Brussels'. *Urban Geography*, 39: 413–37.
Keegan, M. 2018. 'Shenzhen's silent revolution: World's first fully electric bus fleet quietens Chinese megacity'. *The Guardian*, 12 December. Accessed July 2025. www.theguardian.com/cities/2018/dec/12/silence-shenzhen-world-first-electric-bus-fleet.
Keeling, D. 2015. 'A picture is worth 1000 Words: Urban revitalization in an Olympic City: The Perimetral in Rio de Janeiro'. *Focus on Geography*, 58: 194–5.
Keller, R. 2013. *Doing Discourse Research: An Introduction for Social Scientists*. London: Sage.
Kendall, G. and Wickham, G. 1999. *Using Foucault's Methods*. London: Sage.
Kimble, M. 2023. 'The new generation of freeway fighters is assembling'. *Citylab*, 15 November. Accessed July 2025. www.bloomberg.com/news/newsletters/2023-11-15/citylab-daily-the-new-generation-of-freeway-fighters-is-assembling.
Kimmelman, M. 2023. 'How one city tried to solve gridlock for us all'. *New York Times*, 7 December. Accessed August 2024. https://www.nytimes.com/interactive/2023/12/07/headway/bogota-bus-system-transmilenio.html.

King's Cross Central Limited Partnership. 2021. *King's Cross Overview*. London: KCCLP.
King's Cross Central Limited Partnership. 2023. King's Cross – a traffic free oasis. Accessed August 2023. www.kingscross.co.uk/.
Knowles, R. 1996. 'Transport impacts of Greater Manchester's Metrolink light rail system'. *Journal of Transport Geography*, 4: 1–14.
Knowles, R. 2012. 'Transit Oriented Development in Copenhagen, Denmark: From the Finger Plan to Ørestad'. *Journal of Transport Geography*, 22: 251–61.
Kraftl, M. 2021. *Air Quality Management in Oxford City Centre, 1960–2020*. Oxford: Oxfordshire County Council.
Land Transport Authority. 2018. *Public Consultation on Land Transport Master Plan 2040*. Singapore: LTA.
Land Transport Authority. 2019. *Land Transport Masterplan 2040*. Singapore: LTA.
Land Transport Authority. 2023. *North–South Corridor: Redefining journeys for tomorrow*. Singapore: LTA.
Lawrence, M., Bullock, R. and Liu, Z. 2019. *China's High Speed Rail Development*. Washington DC: World Bank. Accessed August 2023. https://openknowledge.worldbank.org/handle/10986/31801.
Leary-Owhin, M. 2016. *Exploring the Production of Urban Space: Differential Space in Three Post-Industrial Cities*. Bristol: Policy Press.
Lefebvre, H. 1947. *Critique of Everyday Life, Volume I: Introduction*. London: Verso (1991).
Lefebvre, H. 1968. 'The right to the city'. In *Writings on Cities (1996)*, edited E. Kofman and E. Lebas. Cambridge, MA: Wiley-Blackwell.
Lefebvre, H. 1974. *The Production of Space*. Paris: Anthropos.
Li, X. and Hu, S. 2016. 'Urban planning history of Shenzhen city since 1979 and its enlightenment'. *Journal of Landscape Research*, 8: 20–2.
Li, Y., Zhan, C., De Jong, M. and Lukszo, Z. 2016. 'Business innovation and government regulation for the promotion of electric vehicle use: Lessons from Shenzhen, China'. *Journal of Cleaner Production*, 134: 371–83.
Liu, Z. and Zhou, L. 2024. 'Central-local fiscal ties in the spotlight'. *China Daily*, 28 August. Accessed September 2024. https://www.chinadaily.com.cn/a/202408/28/WS66ce7cb5a31060630b925513.html.
Living Streets, London Cycle Campaign & Rosehill Highways. 2020. *A Guide to Low Traffic Neighbourhoods*. London: Living Streets, LCC.
Lu, P. 2022. *Developing the High-Quality Dutch Cycling Experience: Lessons from Houten* (unpublished MSc dissertation). Bartlett School of Planning, UCL.
Lucas, K. 2004. *Running on Empty: Transport, social exclusion and environmental justice*. Bristol: Policy Press.
Lucas, K. 2012. 'Transport and social exclusion: Where are we now?' *Transport Policy*, 20: 105–13.
Lyons, G. 2012. 'Visions for the future and the need for a social science perspective in transport studies'. In *Automobility in Transition: A Socio-Technical Analysis of Transport*, edited by F. W. Geels, R. Kemp, G. Dudley and G. Lyons. London: Routledge.
Macharis, C. and Bernardini, A. 2015. 'Reviewing the use of multi-criteria decision analysis for the evaluation of transport projects: Time for a multi-actor approach'. *Transport Policy*, 37: 177–86.
Macharis, C., Turcksin, L. and Lebeau, K. 2012. 'Multi actor multi criteria analysis (MAMCA) as a tool to support sustainable decisions: State of use'. *Decision Support Systems*, 54: 610–20.
Magnusson, J. and Kost, C. 2018. 'Dar es Salaam leads a breakthrough for African cities'. *Sustainable Transport, ITDP*, 29: 6–7.
Malmö City Planning Office. 2011. *The Creative Dialogue for Flagghusen*. Malmö: Malmö City PO.
Malmö City Planning Office. 2015. *Västra Hamnen: Current Urban Development*. Malmö: Malmö City PO.
Marx, K. 1887. *Capital. Volume I: A Critical Analysis of Capitalist Production*. New York: International Publishers (1967).
Maton, K. 2014. Habitus. *Pierre Bourdieu: Key Concepts*. Abingdon: Routledge.
Matthies, E. & Klöckner, C. 2015. 'Car-fixation, socialization, and opportunities for change'. In *Handbook on Transport and Development*, edited by R. Hickman, D. Bonilla, M. Givoni and D. Banister. Cheltenham: Edward Elgar.
Maybee, J. 2020. 'Hegel's dialectics'. In *Stanford Encyclopedia of Philosophy*, edited by E. Zalta. San Francisco: Stanford University.

Mazza, L. and Rydin, Y. 1997. 'Urban sustainability: Discourses, networks and policy tools'. *Progress in Planning*, 47: 1–73.
MBM Arquitectes & AZ Urban Studio. 2003. *Vision for Plymouth*. Plymouth: Plymouth City Council.
McCann, B. 2022. 'The map, the photograph and the BRT: Between promise and reality on Rio's Bus Rapid Transit'. *Cadernos de Arte e Antropologia*, 11: 81–96.
McEwen, V. and Pimental Walker, A. 2015. 'Brazil: Twenty-first century architectures of the mega-event'. *The Avery Review*, 9. Accessed August 2024. https://averyreview.com/issues/9/mega-event.
McVoy, A. D. 1945. 'A history of city planning in Portland, Oregon'. *Oregon Historical Quarterly*, 46: 3–21.
Merriam-Webster Dictionary. 2024. Accessed May 2021. www.merriam-webster.com/.
Metz, D. 2008. 'The myth of travel time saving'. *Transport Reviews*, 28: 321–36.
Ministry of Housing and Urban Affairs. 2017. *National TOD Policy*. Delhi: MOHUA.
Ministry of Transport. 1963. *Traffic in Towns: A Study of the Long Term Problems of Traffic in Urban Areas: Reports of the Steering Group and Working Group appointed by the Minister of Transport [Chairman of Working Group, Sir Colin Buchanan]*. London: HMSO.
Mkalawa, C. and Haixiao, P. 2014. 'Dar es Salaam city temporal growth and its influence on transportation'. *Urban, Planning and Transport Research*, 2: 423–46.
Mladenovic, M. and Trifunovic, A. 2014. 'The shortcomings of the conventional four step travel demand forecasting process'. *Journal of Road and Traffic Engineering*, 60: 5–12.
Mokhtarian, P. L. and Salomon, I. 2001. 'How derived is the demand for travel? Some conceptual and measurement considerations'. *Transportation Research Part A: Policy and Practice*, 35: 695–719.
Monheim, R. 1990. 'The evolution and impact of pedestrian areas in the Federal Republic of West Germany'. In *The Greening of Urban Transport: Planning for walking and cycling in western cities*, edited by R. Tolley. London: Belhaven Press.
Montezuma, R. 2005. 'The transformation of Bogotá, Colombia, 1995–2000: Investing in citizenship and urban mobility'. *Global Urban Development*, 1.
Moscoso, M., Van Laake, T. and Quiñones, L. 2019. *Sustainable Urban Transport in Latin America: Assessment and recommendations for mobility policies*. Bogotá: Despacio.
Mumford, L. 1963. *The Highway and the City*. New York: Harvest Books.
National Capital Region Transport Corporation. 2020. *Delhi–Meerut Regional Rapid Transit System Investment Project: EIA*. Delhi: NCRTC, ADB.
National Capital Region Transport Corporation. 2024. *Regional Rapid Transit System (RRTS): Implementation of Transit Oriented Development (TOD)*. Delhi: NCRTC.
Nello-Deakin, S. and Nikolaeva, A. 2021. 'The human infrastructure of a cycling city: Amsterdam through the eyes of international newcomers'. *Urban Geography*, 42: 289–311.
Netherlands Institute for Infrastructure and Water. 2018. *Mobility Picture and Key Figures*. Amsterdam: Knowledge Institute for Mobility Policy.
Netherlands Institute for Transport Policy. 2018. *Cycling Facts*. The Hague: Ministry of Infrastructure and Water Management.
Newman, P. and Kenworthy, J. 2015. *The End of Automobile Dependence: How cities are moving beyond car-based planning*. Washington DC: Island Press.
No More Freeways. 2024. 'About No More Freeways. *No More Freeways*. Accessed April 2024. https://nomorefreewayspdx.com/.
Nordbakke, S. 2013. 'Capabilities for mobility among urban older women: Barriers, strategies and options'. *Journal of Transport Geography*, 26: 166–74.
Norton, P. 2011. *Fighting Traffic: The dawn of the motor age in the American city*. Cambridge, MA: MIT Press.
O'Farrell, C. 2005. *Michel Foucault*. London: Sage.
Office for National Statistics. 2011. *2011 Census*.
Oldenziel, R. and Albert de la Bruhèze, A. 2011. 'Contested spaces: Bicycle lanes in urban Europe, 1900–1995'. *Transfers*, 1: 29–49.
Oosterhuis, H. 2016. 'Cycling, modernity and national culture'. *Social History*, 41: 233–48.
Oregon Department of Transportation. 2022. *Urban Mobility Strategy*. Portland: Urban Mobility Office.
Oregon Department of Transportation. 2023. *Oregon Transportation Plan*. Portland: Oregon Transportation Commission.

Ortega, S. 2015. '"Chaos? This is open-heart surgery": Medellín risks a massively expensive plan to bury its highway'. *The Guardian*, 4 May. Accessed July 2025. www.theguardian.com/cities/2015/may/01/medellin-bury-highway-urban-intervention.

Oxford City Council. 1973. *A Balanced Transport Policy: Report of Central Area Working Party*. Oxford: OCC.

Oxfordshire County Council. 2015. *Oxford Transport Strategy*. Oxford: OCC.

Oxfordshire County Council. 2019. *Connecting Oxford*. Oxford: OCC.

Paget-Seekins, L. 2015. 'Bus rapid transit as a neoliberal contradiction'. *Journal of Transport Geography*, 48: 115–20.

Parkhurst, G. 1995. 'Park and ride: Could it lead to an increase in car traffic?' *Transport Policy*, 2: 15–23.

Paterson, M. 2007. *Automobile Politics: Ecology and cultural political economy*. Cambridge: Cambridge University Press.

Peck, J. and Theodore, N. 2010. 'Mobilizing policy: Models, methods, and mutations'. *Geoforum*, 41: 169–74.

Peck, J. and Ward, K. 2002. 'Placing Manchester'. In *City of Revolution: Restructuring Manchester*, edited by J. Peck and K. Ward. Manchester: Manchester University Press.

Peñalosa, E. 2024. *Equality and the City: Urban Innovations for All Citizens*. Philadelphia: University of Pennsylvania Press.

Pérez, J. 2018. 'The dystopian, utopian parable of Medellín'. *Kings Review*, May.

Perry, C. 1929. 'The neighbourhood unit: A scheme for arrangement for the family-life community'. *Regional Survey of New York and Its Environs*. New York: Arno Press (1974).

Pharoah, T. 1992. *Less Traffic, Better Towns*. London: Friends of the Earth.

Plymouth City Council. 2014. *The Plymouth Report*. Plymouth: PCC.

Prince's Foundation. 2004. *Sherford New Comunity Enquiry by Design: Summary report*. London: The Prince's Foundation.

Pucher, J. and Buehler, R. 2008. 'Making cycling irresistible: Lessons from The Netherlands, Denmark and Germany'. *Transport Reviews*, 28: 495–528.

Rajé, F. 2007. 'Using Q methodology to develop more perceptive insights on transport and social inclusion'. *Transport Policy*, 14: 467–77.

Rapid Transition Alliance. 2018. *The Medellín Miracle*. Brighton: Institute of Development Studies. Accessed June 2025. www.rapidtransition.org/stories/the-medellin-miracle/.

Ren, D. 2018. 'Shenzhen's all-electric bus fleet is a world's first that comes with massive government funding'. *South China Morning Post*, 23 October. Accessed August 2024. https://www.scmp.com/business/china-business/article/2169709/shenzhens-all-electric-bus-fleet-worlds-first-comes-massive.

Robinson, J. 2015. 'Arriving at' urban policies: The topological spaces of urban policy mobility'. *International Journal of Urban and Regional Research*, 39: 831–4.

Rose, M. and Rohl, R. 2012. *Interstate: Highway politics and policy since 1939*. Knoxville: University of Tennessee Press.

Ryan, J., Wrestrand, A. and Schmidt, S. M. 2015. 'Exploring public transport as an element of older persons' mobility: A capability approach perspective'. *Journal of Transport Geography*, 48: 105–14.

Said, E. 1978. *Orientalism*. London: Routledge and Kegan Paul.

Salford City Council. 2005. *Salford Quays Milestones: The Story of Salford Quays*. Salford: SCC.

Schwanen, T., Dijst, M. and Dieleman, F. 2004. 'Policies for urban form and their impact on travel: the Netherlands experience'. *Urban Studies*, 41: 579–603.

Schwanen, T., Lucas, K., Akyelken, N., Cisternas Solsona, D., Carrasco, J.-A. and Neutens, T. 2015. 'Rethinking the links between social exclusion and transport disadvantage through the lens of social capital'. *Transportation Research, Part A*, 74: 123–35.

Science and Industry Museum. 2021. 'Manchester's smoke nuisance: Air pollution in the industrial city'. Manchester. Accessed December 2021. www.scienceandindustrymuseum.org.uk/objects-and-stories/our-environment/air-pollution.

Secretaria Distrital de Movilidad Bogotá. 2019. *Bogotá Capital Mundial de la Bici: Una Visión de Ciudad*. Bogotá: Secretaria Distrital de Movilidad, Mayor of Bogotá.

Sen, A. 1985. *Commodities and Capabilities*. Amsterdam: North-Holland.

Sen, A. 1999. *Development as Freedom*. Oxford: Oxford University Press.

Sen, A. 2009. *The Idea of Justice*. London: Allen Lane.

Sharp, L. and Richardson, T. 2001. 'Reflections on Foucauldian discourse analysis in planning and environmental policy research'. *Journal of Environmental Policy & Planning*, 3: 193–209.
Sheller, M. 2004. 'Automotive emotions: feeling the car'. *Theory, Culture & Society*, 21: 221–42.
Sheller, M. and Urry, J. 2000. 'The city and the car'. *International Journal of Urban and Regional Research*, 24: 737–57.
Shepley, C. 1991. 'Planning Plymouth's future'. In *Plymouth: Maritime City in Transition*, edited by B. Chalkley, D. Dunkerley and P. Gripaios. London: David & Charles.
Sherford Building Futures. 2018. Sherford Masterplan. Accessed November 2021. https://sherford.org/.
Shields, R. 1999. *Lefebvre, Love and Struggle*. London: Routledge.
Shliselberg, R. and Givoni, M. 2018. 'Motility as a policy objective'. *Transport Reviews*, 38: 279–97.
Shore, S. 2023. 'How does the presence of light rail impact social equity across the city? A qualitative study of Sheffield, England through the motility lens' (unpublished MSc dissertation). UCL.
Shortall, R. 2021. 'Chapter Three – Deliberative appraisal methods'. In *Advances in Transport Policy and Planning*, edited by N. Mouter. Academic Press.
Shumway, D. 1989. *Michel Foucault*. Charlottesville: University Press of Virginia.
Silverman, D. 2013. *Doing Qualitative Research*. London: Sage.
Simmie, J. 1974. *Citizens in Conflict: The Sociology of Town Planning*. London: Hutchinson.
Singapore Government. 2023a. *Key Household Income Trends*. Singapore: Department of Statistics.
Singapore Government. 2023b. *Summary Table: Income*. Singapore: Ministry of Manpower.
Smith, N. 1987. 'Gentrification and th rent gap'. *Annals of the Association of American Geographers*, 77: 462–5.
Smith, N. 2002. 'New globalism, new urbanism: Gentrification as global urban strategy'. *Antipode*, 34: 427–50.
Smoak, S. 2007. *Framing the Automobile in Twentieth Century American Literature: A spatial approach*. University of North Carolina.
Social Exclusion Unit. 2003. *Making the Connections: Final report on transport and social exclusion*. London: SEU/ODPM.
Statista. 2016. 'Statistics and facts about automobile advertising in the US'. Accessed June 2026. www.statista.com/topics/1601/automotive-advertising/.
Statistics Netherlands. 2022. 'Houten population estimate'. Accessed June 2022. www.citypopulation.de/en/netherlands/admin/utrecht/0321__houten/.
Steer. 2021. *London Borough of Ealing Low Traffic Neighbourhood Consultation Analysis*. London: Steer, Ealing Council.
Swyngedouw, E. 2007. 'Impossible 'sustainability' and the post-political condition'. In *The Sustainable Development Paradox*, edited by D. Gibbs and R. Krueger. New York: Guilford Press.
Swyngedouw, E. and Kaika, M. 2014. 'Urban political ecology: Great promises, deadlock ... and new beginnings'. *Documents d'Anàlisi Geogràfica*, 60: 459–81.
Swyngedouw, E., Moulaert, F. and Rodriguez, A. 2002. 'Neoliberal urbanization in Europe: Large-scale urban development projects and the new urban policy'. *Antipode*, 34: 542–77.
Transport for Greater Manchester. 2017. *Greater Manchester Transport Strategy 2040: Evidence Base*. Manchester: TfGM.
Transport for Greater Manchester. 2021. *Transport Strategy 2040*. Manchester: TfGM, GMCA, GMLEP.
Transport for London. 2018. *Mayor's Transport Strategy*. London: GLA, TfL.
Tripp, H. A. 1936. 'The traffic problem'. *The Police Journal*, 9: 74–97.
Tripp, H. A. 1938. *Road Traffic and its Control*. London: Edward Arnold.
Tyson, W. 2004. 'Manchester Metrolink Tram System'. *Japan Railway & Transport Review*, 38.
UN-Habitat. 2019. *The Story of Shenzhen: Its economic, social and environmental transformation*. Nairobi: United Nations Human Settlements Programme.
United Nations Census Bureau. 2020. Portland City, Oregon. Accessed May 2024. https://data.census.gov/table/DECENNIALDHC2020.P1?q=portland%20oregon.
United Nations Department of Social and Economic Affairs. 2014. *World Urbanization Prospects: The 2014 Revision*. New York: UN.
United Nations Department of Social and Economic Affairs. 2015. *World Population Prospects: The 2015 Revision*. New York: UN.
Urban Redevelopment Authority. 2019. *Masterplan*. Singapore: URA.

Urban Redevelopment Authority. 2022. *Long Term Plan Review 2021: Space for our dreams*. Singapore: URA.
Urban Task Force. 1999. *Towards an Urban Renaissance*. London: E&F Spon.
Urry, J. 2004. 'The system of automobility'. *Theory, Culture & Society*, 21: 25–39.
US Department of Transportation. 2021a. 'Annual vehicle distance traveled in miles and related data by highway category and vehicle type 2020'. Federal Highway Administration. Accessed April 2024. www.fhwa.dot.gov/policyinformation/statistics/2020/vm1.cfm.
US Department of Transportation. 2021b. 'Public road length – 2020 miles by functional system'. Federal Highway Administration. Accessed April 2024. www.fhwa.dot.gov/policyinformation/statistics/2020/hm20.cfm.
Van Dijk, T. A. 1993. 'Principles of critical discourse analysis'. *Discourse & Society*, 4: 249–83.
Villela, G. 2014. 'Perimetral Elevator was inaugurated twice: By JK, in 1960, and Geisel, in 1978'. *O Globo*, 16 July. Accessed February 2020. https://acervo.oglobo.globo.com/em-destaque/elevado-da-perimetral-foi-inaugurado-duas-vezes-por-jk-em-1960-geisel-em-1978-13279735.
Walker, J. 2010. 'Portland: Another challenging chart'. *Human Transit*. Accessed May 2024. https://humantransit.org/2010/01/portland-another-challenging-chart.html.
Weber, R. 2002. 'Extracting value from the city: Neoliberalism and urban redevelopment'. *Antipode*, 34: 519–40.
Whyte, W. H. 1980. *The Social Life Of Small Urban Spaces*. New York: Project for Public Spaces.
Wodak, R. 2001. 'What CDA is about – a summary of its history, important concepts and its developments'. In *Methods of Critical Discourse Analysis*, edited by R. Wodak and M. Meyer. London: Sage.
Wodak, R. and Meyer, M. 2001. *Methods of Critical Discourse Analysis, Volume 1: Concepts, history, theory*. London: Sage.
Woodcock, J., Edwards, P., Tonne, C., Armstrong, B. G., Ashiru, O., Banister, D., Beevers, S., Chalabi, Z., Chowdhury, Z., Cohen, A., Franco, O. H., Haines, A., Hickman, R., Lindsay, G., Mittal, I., Mohan, D., Tiwari, G., Woodward, A. and Roberts, I. 2009. 'Public health benefits of strategies to reduce greenhouse-gas emissions: Urban land transport'. *The Lancet*, 374: 1930–43.
World Bank. 2022. *World Development Indicators*. Washington DC: World Bank.
World Health Organization. 2015. *Noncommunicable Diseases Fact Sheet*. Geneva: WHO. Accessed June 2025. www.who.int/mediacentre/factsheets/fs355/en/.
World Health Organization. 2023. *Global Status Report on Road Safety*. Geneva: WHO.
Wozniacka, G. 2024. 'Legal action over Metro's transportation plan brings climate challenges to forefront'. *The Oregonian*, 13 February. Accessed August 2024. https://www.oregonlive.com/environment/2024/02/legal-action-over-metros-transportation-plan-brings-climate-planning-challenges-to-forefront.html.
Wu, F. 2022. *Creating Chinese Urbanism: Urban revolution and governance changes*. London: UCL Press.
Wu, F., Zhang, F. and Liu, Y. 2022. 'Beyond growth machine politics: Understanding state politics and national political mandates in China's urban redevelopment'. *Antipode*, 54: 608–28.

# Index

ABC location policy 124
Abercrombie, P. 55, 63
access 8, 24, 35, 36, 56, 72, 136, 148, 157, 158, 160, 192, 210, 221, 222, 223, 224, 226, 229, 232, 238, 241, 246, 248, 249, 250, 254, 259, 265, 266, 267, 271, 272, 274, 277, 293, 294, 302, 309, 310, 311, 312, 318, 319, 320
apparatus 33, 48, 62, 71, 100, 123, 134, 136, 152, 153, 164, 244, 291, 294, 295, 298, 304
appropriation 25, 35, 36, 63, 84, 222, 224, 238, 244, 254, 298, 307, 309, 310, 311, 312
archaeology 32, 43, 55
avoid-shift-improve framework 77

Beaux-Arts 55, 302
biopower 33, 84, 134, 155, 164, 291, 295, 304, 305
Bogotá 13, 39, 72, 111–21, 158, 250, 254, 303, 310, 311, 320
   TransMilenio 112

Capabilities Approach 222
capital 36
Chongqing 13, 39, 135–42, 304
cities 8, 18
city planning 18
climate change 3, 5, 6, 7, 49, 65, 110, 180, 266, 276, 292, 298, 300, 316, 318
competence 35, 36, 222, 309, 310, 311, 312
conceived space 37, 55, 128, 259
conduct 34, 133, 134, 256, 266, 271, 291, 296, 308, 320
congestion 4
   congested 11
contestation 17, 18, 21, 22, 25, 37, 99, 134, 255, 273, 274, 280, 296, 300, 303, 308, 314
convenient 17
Copenhagen 13, 39, 68, 167, 179–88, 190, 305
CROW design manual 126
Cubitt, L. 148
culture 5, 10, 32, 34, 48, 49, 56, 120, 173, 183, 222, 256, 267, 291, 294, 295, 296, 297, 302, 308
Curitiba 112, 158, 200
cycling 3, 8, 12, 21, 90, 100, 116, 120, 123, 132, 135, 166, 169, 170, 172, 173, 179, 180, 181, 182, 183, 191, 192, 210, 223, 232, 233, 246, 257, 258, 259, 260, 265, 266, 267, 273, 275, 278, 279, 292, 294, 298, 299, 302, 303, 304, 305, 306, 308, 309, 310, 311, 320

Dar es Salaam 13, 39, 167, 199–205, 306
deductive 41
Delhi 13, 39, 56, 167, 207–13, 306
deliberative 5, 13, 268, 314, 316, 317
democratic 17
discontinuity 13, 26, 28, 33, 39, 42, 43, 165, 167, 180, 200, 256, 291, 295, 296, 301, 305, 312, 315, 316
discourse 13, 27, 28, 29, 32, 33, 84, 87, 102, 105, 132, 166, 263, 291, 294, 297, 298, 299, 300, 301, 302, 303, 312, 313, 314, 316, 319, 321
   discourses 12, 13, 28, 33, 48, 72, 279, 292, 293, 294, 295, 296, 297, 312, 317, 320, 321
   discursive struggle 17, 18
discursive formation 31, 46, 48, 54, 55, 99, 112, 132, 191, 276, 291, 294, 295, 299, 300, 301
discursive meaning 32, 46, 67, 295
discursive practice 32, 62, 66, 85, 192, 244, 295, 300, 304
doxa 16

efficiency 24
efficient 17
energy consumption 7, 18, 24, 49, 52, 54, 210, 299
environmental movement 87, 273
episteme 32, 43, 48, 56, 72, 85, 99, 110, 115, 291, 294, 295, 301, 302, 303
ethics 33, 39, 221, 254, 291, 296, 299, 301, 307, 312, 315
   social equality 221
   social equity 221
   social justice 221
event 32, 33, 43, 55, 57, 120, 134, 165, 291, 295, 302, 303
exclusion 34, 152, 154, 222, 245, 256, 267, 291, 294, 296, 301, 307, 308
experience 19, 33, 35, 43, 61, 165, 173, 179, 192, 200, 218, 221, 222, 254, 259, 262, 291, 294, 296, 307, 320

335

Foucault, M. 5, 6, 26, 29, 30, 71, 133, 221, 301, 321
    Foucauldian 30, 38, 55, 165
freedom 18, 46, 173, 255
freeway removal 273
Freiburg 13, 39, 68, 72, 87–95, 303, 310, 315
    Rieselfeld 95, 303, 310
    Vauban 95, 303, 310

genealogy 26, 33, 43, 55, 134
Global North 39
Global South 39
governmentality 33, 134, 138, 144, 271, 295, 304

health 7, 8, 29, 49, 53, 54, 56, 154, 158, 180, 222, 260, 299, 310
    inactivity 6, 7, 53
    obesity 6, 7, 53, 172
hegemony 19, 35, 265
high-speed rail 110, 134, 135, 138, 215, 232, 304
highway planning 39
history 5, 29, 30, 32, 39, 43, 99, 120, 166, 167, 179, 181, 199, 225, 229, 238, 255, 263, 269, 270, 274, 281, 282, 291, 295, 296, 299, 301, 302, 303, 308, 312, 315, 320, 321
Houten 13, 39, 41, 72, 123–32, 304, 310, 311, 320
    Houten Castellum 125

ideals 17
informal neighbourhoods 157, 305
informal transport 3
institutionally racist planning 270
interpretative 25
interstate 18, 271, 272, 274, 275, 278

jaywalking 18, 64, 104

Kaufmann, V. 35, 36, 222, 309, 319
knowledge 4, 8, 26, 28, 32, 33, 34, 35, 36, 43, 65, 71, 72, 134, 154, 155, 166, 191, 223, 276, 280, 282, 291, 292, 294, 295, 296, 297, 300, 302, 303, 304, 307, 309, 315, 320

Lefebvre, H. 36, 37, 256, 259, 293
liberator 17
lived space 37, 128, 259, 262
London, Ealing LTN21 13, 39, 41, 257–67, 308
London, King's Cross 13, 39, 145–52, 304
low traffic neighbourhood 10, 19, 256, 257, 258, 308

Malmö 13, 39, 68, 167, 189–98, 306, 315
    Västra Hamnen 189, 192, 306
Manchester 13, 39, 224, 225–37, 307
    Castlefield 229
    Metrolink 226
Medellín 13, 39, 116, 224, 245–53, 307, 315
    cable car 248
modern 17
modernity 19

motility 35, 36, 84, 210, 309, 310, 311, 312, 319
motor lobby 17
motordom 18
motorisation 3, 7, 12, 19, 40, 43, 45, 46, 55, 65, 66, 111, 123, 165, 166, 169, 223, 225, 226, 229, 265, 271, 272, 274, 277, 282, 291, 292, 293, 296, 299, 300, 302, 308, 320, 321

naturalised 15, 19, 301
neoliberal 17, 46, 48, 114, 115, 122, 152, 293, 298
    neoliberalism 19, 48, 64, 122, 152, 154, 311, 312
    neoliberalist 255
No More Freeways 276, 280
non-communicable diseases 6, 53
    cancers 53
    cardiovascular diseases 53
    diabetes 53
    respiratory diseases 53
normalisation 33, 71, 115, 122, 291, 295, 303

Opportunity Areas 153
OV-fiets 127, 170
Oxford 13, 19, 39, 72, 75–82, 257, 303
    Christ Church meadow 76

participation 4, 5, 40, 65, 87, 99, 143, 158, 164, 210, 222, 223, 238, 245, 246, 266, 272, 294, 303, 306, 307, 308, 312, 316, 317, 319, 320
    participatory 5, 13, 56, 97, 153, 246, 254, 263, 268, 312, 313, 314, 316, 317
Pearl River Delta 215
pedestrianisation 58, 78, 88, 303, 309
Peñalosa, E. 112
perceived space 37, 259, 260
Plymouth 13, 39, 43, 55–62, 302, 309
political xv, 3, 4, 5, 10, 13, 17, 19, 21, 22, 26, 31, 87, 154, 180, 222, 225, 241, 246, 247, 258, 269, 278, 282, 293, 296, 307, 315, 316, 318, 319, 320, 321
Portland 13, 39, 56, 256, 269–87, 308
    Albina 270, 274
power 4, 32, 33, 34, 39, 43, 71, 87, 133, 134, 136, 145, 152, 153, 154, 158, 164, 223, 226, 255, 256, 276, 282, 291, 292, 294, 295, 296, 299, 300, 301, 303, 304, 312, 314, 315, 317, 319, 320
practice 4, 15, 16, 27, 28, 31, 33, 35, 37, 39, 43, 55, 63, 64, 65, 153, 158, 165, 166, 172, 180, 192, 218, 232, 257, 265, 270, 291, 292, 294, 295, 296, 301, 304, 305, 306, 310, 318
problematisation 8, 25, 30, 33, 65, 97, 166, 169, 218, 291, 295, 297, 305
progressive 17
public 13
public policy 3, 114, 153, 154, 165, 179, 180, 192, 210, 238, 241, 244, 248, 255, 275, 278, 279, 292, 293, 296, 297, 298, 299, 300, 301, 309, 312, 313, 314, 315, 316, 317, 320

public transport 3, 8, 12, 41, 77, 90, 100, 102, 112, 123, 135, 148, 152, 157, 158, 164, 166, 167, 169, 170, 172, 181, 182, 199, 200, 201, 210, 215, 216, 217, 218, 222, 223, 228, 229, 230, 231, 232, 233, 238, 239, 241, 250, 254, 258, 265, 266, 269, 271, 272, 273, 274, 275, 278, 279, 292, 293, 298, 299, 300, 302, 303, 304, 305, 306, 308, 309, 310, 311, 320

rational 11
regeneration 154
regime of truth 84, 87, 102, 114, 263, 291, 295, 303
resistance 133
Rio de Janeiro 13, 39, 157–63, 254, 305
   Porto Maravilha 134, 158, 305

Shenzhen 13, 39, 167, 215, 306
   electric vehicles 216
Singapore 13, 39, 72, 99–105, 303, 310
   Concept Plan 100
   electronic road pricing 101
   Land Transport Authority 100
   Urban Redevelopment Authority 100
spatial triad 37, 259
special economic zone 215
streetcar 18, 88, 111
streetspace 19
subject 4, 28, 32, 33, 34, 71, 104, 210, 223, 232, 256, 270, 291, 294, 296, 303, 308
subjectivity 25, 34, 39, 46, 255, 256, 291, 296, 299, 301, 307, 308, 312, 315
sustainability 23

techne 32, 72, 83, 85, 87, 126, 291, 295, 303, 304
traffic casualties 6, 7, 18, 19, 24, 49, 52, 54, 62, 123, 210, 299
traffic engineering 18, 46, 62, 63, 293
   traffic engineers 21, 48, 83, 265

traffic filtering 96, 126
transit-orientated development 20, 145, 153, 155, 210, 304
   transit-orientated design 137
   transit-orientated neighbourhood 134
transport planning 11, 39
transport system 3, 4, 158, 199, 200, 201, 210, 215, 222, 223, 229, 231, 232, 250, 254, 298, 300, 305, 306, 307, 309, 310, 312, 313, 318, 319, 320
   transport systems 3, 4, 12, 13, 157, 158, 166, 169, 201, 222, 223, 225, 229, 256, 278, 291, 292, 293, 294, 295, 296, 297, 298, 299, 300, 301, 307, 311, 312, 316, 319, 320, 321
travel behaviours 3, 4, 5, 9, 12, 167, 169, 200, 255, 257, 261, 263, 269, 272, 291, 292, 294, 296, 297, 298, 299, 301, 309, 312, 313, 316, 318, 319, 320, 321
Tripp, A. 62, 63
truth 28, 32, 33, 39, 43, 46, 71, 126, 133, 255, 291, 295, 296, 297, 299, 301, 303, 312, 315, 320, 321

urban flyovers 19
Utrecht 13, 39, 68, 124, 167, 169–77, 305, 311, 320

Valenciennes 13, 39, 224, 239–44, 307
value 26
VINEX strategy 124

walking 3, 8, 166, 169, 170, 172, 181, 191, 199, 210, 223, 246, 257, 259, 260, 265, 273, 274, 275, 278, 279, 292, 293, 294, 298, 299, 302, 306, 310, 311, 320
World Bank 13, 113, 114, 200
World City 153

Yangtze River 136